P9-DYD-758

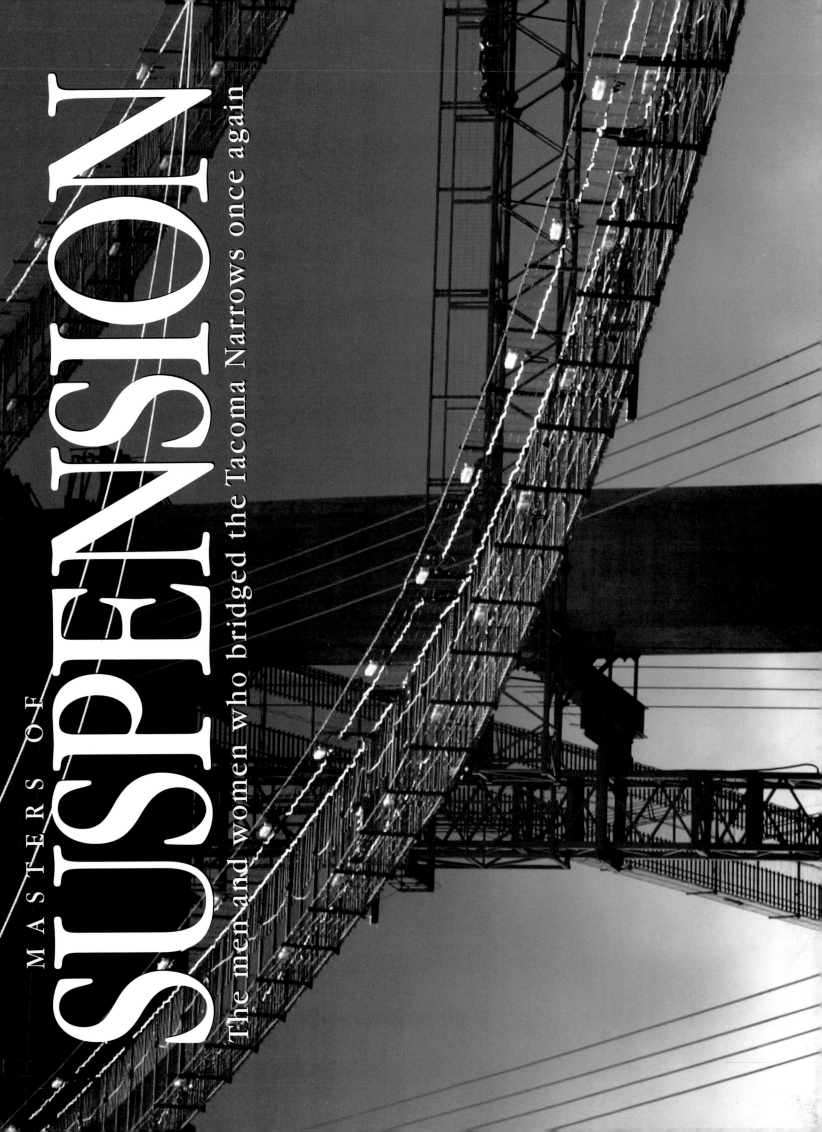

MASTERS OF SUSPENSION

The men and women who bridged the Tacoma Narrows once again

Author **ROB CARSON**

Photographer **DEAN J. KOEPFLER**

Editor **RANDALL MCCARTHY**

Photo editor **JEREMY HARRISON**

Graphic artist **FRED MATAMOROS**

CONTENTS

GETTING BEYOND GERTIE

With one bridge a flamboyant failure and another overwhelmed by traffic, a community struggles unsteadily toward a third crossing.

Shortly before noon on Wednesday, March 8, 2000, just as Marvin Parkko drove onto the east end of the Tacoma Narrows Bridge, he felt the left front wheel come off his 1964 Kenworth logging truck.

Parkko's truck lurched to the left, across the centerline of the bridge and into the path of an oncoming propane tanker heading back from a delivery in Port Townsend. The tanker driver swerved to avoid Parkko and suddenly there was chaos.

The tanker hit a red pickup truck with a glancing blow, setting it spinning in the middle of the bridge. It sideswiped two more cars, then plowed over a third, a blue Chevrolet, crushing it and its driver against the concrete bridge rail.

Collisions moved backward in both directions as drivers hit their brakes. In seconds, all traffic on the mile-long bridge had stopped. To the east, cars and trucks jerked to a halt all across Tacoma's industrial Nalley Valley, and to the west, they

▲ By 2000, the Tacoma Narrows Bridge was overwhelmed with traffic—32 million cars and trucks crossed it that year. Narrow lanes and the lack of a median barrier made it one of the most dangerous stretches of highway in Washington state. Between 1990 and 2000, 11 people died on the bridge or its approaches.

▲ "Galloping Gertie," under construction in 1939 at the Tacoma Narrows, a mile-wide passage noted for its high winds and rapid currents.

backed up through Gig Harbor, past the women's prison at Purdy and, eventually, halfway to Port Orchard, 18 miles away.

In the Chevrolet, Kurt Ferguson, a 36-year-old father of two young children, was dead. As drivers wondered how to get him out, his car burst into flames. Police evacuated nearby houses, afraid the propane tanker might blow up.

Crashes were not unusual on the Narrows bridge in 2000. The steel bridge was so narrow and so crowded, a single miscalculation could start a demolition derby. On one foggy morning in February 1992, there had been seven accidents on the bridge in two hours. Sixteen cars collided, injuring 12 people. Between 1993 and 1996, there were 513 injury accidents on the bridge and its approaches. Ferguson was the eighth person to die in as many years.

The bridge turned 50 years old in 2000 and earned a place on the National Register of Historic Places. But by then it had become so overwhelmed with traffic that many people went to great lengths to avoid it. If crossing was unavoidable, they did so in fear, hands clenched to the wheel and hugging the outside curb. Daily traffic jams on the bridge left commuters fuming and routinely backed up traffic into Tacoma's residential streets.

It was clear in 2000 that something needed to be done about the bottleneck at the Narrows, but arguments about solutions had raged for 10 years without result.

Parkko's left front wheel, and the particularly horrible accident it set in motion, marked a turning point in the controversy. The momentum changed. Arguments against building a new bridge began sounding hollow, lawsuits seemed petty. Talk of building a tunnel under the Narrows, of double decking the existing bridge or building long-span bridges that hopped from island to island elsewhere in Puget Sound all died down.

Attention focused on what appeared to be the most obvious and practical solution—a new suspension bridge across the Narrows, parallel to the existing span.

Bill Matheny, an 86-year-old ironworker who had helped build not only the 1950 bridge but also its predecessor, put the feelings of impatience into words.

"Nobody likes to pay taxes," he said, "but if you're going to move from Point A to Point B, you've got to have a way to do it, don't you? Let's spit on our gloves and get the damn job done."

The cost of a new bridge would be huge. It would be one of the state's largest single public works project, the longest suspension bridge built in the United States in nearly 40 years. The cost, estimated at $850 million, would be as much as NASA's Mars Rover project.

What's more, the new bridge would have to be built in a location notorious for failure. The Narrows, a natural funnel for high winds and tides, was the site of one of the worst engineering disasters in history. The first bridge across the passage, nicknamed Galloping Gertie, shook itself apart in a 1940 windstorm.

Still, geography made a strong argument for locating the new bridge at the Narrows, just as it had more than 100 years earlier when pioneers first envisioned a crossing there. Puget Sound cuts down into Washington state from Canada in a big "V," with the Olympic Peninsula jutting out from the rest of the state like a hitchhiker's thumb. There is no way to get across the Sound by land, short of driving to the base of the V at the state capital in Olympia and heading up the other side.

In most places the distance across the Sound is too great to consider bridge building. But at the

Tacoma Narrows, peninsula and mainland are tantalizingly close, just more than a mile apart, with high, stable banks on both sides.

Pioneers predicted a bridge there as early as 1889. Even then it was clear the site would have its challenges. Because the channel is so narrow, it restricts and therefore accelerates wind and tides. Between tidal exchanges, when the water was flat and calm, it was no big deal for settlers to row across in a rowboat. But with the tide running, the Narrows worked like a high-pressure hose, producing currents

▶ **Excavating for Gertie's anchorage on the Tacoma side of the channel in 1939. Workers filled the pit with 25,000 cubic yards of concrete to anchor the ends of the mile-long main cables.**

▲ **The population of the Gig Harbor area tripled between 1980 and 2000. Spectacular views from the bridge contributed to traffic jams.**

strong enough to send boats surfing toward Seattle or Olympia.

The water barrier created a cultural divide as well as a geographic one. The east side of the Sound had the railroads, the smokestacks, the highways. The west side, relatively inaccessible, had forests, Indian reservations and wilderness. Civic boosters in struggling peninsula towns such as Port Angeles, Port Townsend and Bremerton complained about the lack of access, but plenty of west-siders preferred being off the beaten track, tending their chickens and vegetable gardens, clear of the hubbub on the other side.

Nowhere was the cultural divide more obvious than at the Narrows. On the east side, transcontinental railroad connections had turned Tacoma's Commencement Bay into a hive of industrial activity. The Asarco copper smelter blasted a plume of arsenic and lead from its 520-foot stack, carrying a pall eastward toward the Cascades.

Meanwhile, across the water, nonconformists and individualists celebrated the simple life. At Carr Inlet, 12 miles by boat from the Narrows, free thinkers founded a utopian commune at Home in 1896, advocating free love, free speech and women's rights.

Theosophists and socialists gathered at Burley Colony at the head of Henderson Bay, six miles by wagon from the Narrows. They promoted universal brotherhood, equal pay for all jobs and used as a motto: "Make way for brotherhood, make way for man."

Before the first bridge was built, crossing from one side to the other meant taking a ferry from Gig Harbor to docks at the foot of Tacoma's Sixth Avenue or Point Defiance. That was enough of a connection for many peninsula residents.

> "Make way for brotherhood, make way for man."
>
> —BURLEY COLONY MOTTO

But backers of commerce were relentless, and in the 1930s they effectively used national defense to support their argument that the federal government should help finance a bridge. With war appearing likely in Europe, they argued, a connection between the Army's Fort Lewis in Tacoma and the Navy's shipyard at Bremerton was critical.

At that time, a suspension bridge was a logical choice of bridge designs for the Narrows. Long-span suspension bridges were back in favor after several spectacular aerodynamic failures in the previous century. Windstorms took out three suspension bridges in the United States and seven in Europe during the 1800s. But three New York bridges—the rock-solid Brooklyn Bridge, built in 1883, the 1909 Manhattan Bridge and the 1931 George Washington Bridge—changed people's minds.

Those bridges were so stable in the wind that designers began trimming down plans for heavy towers and supporting trusses, believing them unnecessarily clunky.

The most persuasive advocate for lighter, more elegant suspension bridges was Leon Moisseiff, a Latvian emigrant who designed the Manhattan Bridge. With that bridge, Moisseiff applied what he called the "deflection theory" to suspension bridge design.

Moisseiff believed that much of the force of wind blowing against a suspension bridge could be absorbed by the cable system and transmitted to the towers and anchorages. If a bridge were designed to sway moderately back and forth with the winds instead of being completely rigid, he reasoned, the deck would act as a counterweight and help restore equilibrium.

▲ View of what would become Highway 16 from the top of the Tacoma tower in January 1940 as crews prepared to spin Gertie's main cables. At the time, western Tacoma was mostly logged-over forests and open fields.

▲ Rare 1940 color shot of Gertie showing the slim deck profile advocated by Leon Moisseiff. It was more graceful and $4 million cheaper than Clark Eldridge's design.

▶ Gertie's deck was supported by steel girders just 8 feet high. Even during construction, the deck moved so much in the wind it made workers seasick.

Moisseiff argued persuasively that suspension bridges could be built much lighter and still retain their strength. During the Great Depression, those were convincing arguments. Lighter bridges meant less steel and therefore less expense.

The achingly beautiful Golden Gate Bridge, built in 1937 on the principles of the deflection theory, cemented Moisseiff's reputation. The San Francisco main span, at that point the longest in the world at 4,200 feet, was also the most slender and flexible. It had an unprecedented width-to-length ratio of 1 to 47.

The year the Golden Gate Bridge was finished, the Washington Department of Highways asked its chief bridge engineer, Clark Eldridge, to design a similar span across the Narrows. Eldridge knew Moisseiff's work, but he took a conservative approach nonetheless. His design called for a deck stiffened by a framework of 25-foot-deep trusses beneath it. The financing request sent to the federal Public Works Administration came to $11 million.

During its deliberations, the agency also heard from Moisseiff, who advised that the Narrows bridge could be built more simply and inexpensively with simple flat girder sides and no supporting trusses. It would be a better looking bridge, too, he said, looking less like a railroad trestle and more like a long, graceful ribbon across the Narrows. His design also would save more than $4 million.

Moisseiff's reputation cowed the PWA. The agency agreed to finance loans for $6.4 million rather than $11 million. It strongly suggested the Washington Department of Highways bring Moisseiff into the project as a consultant.

Moisseiff reworked Eldridge's plan, taking the deflection theory to new extremes. The Narrows bridge would have a simple plate girder stiffening system just 8 feet deep, with no trusses below. The ratio of its width to length would be 1 to 72, making it significantly slimmer than the Golden Gate. The deck structure would weigh only about a tenth as much as any other major suspension bridge.

The Narrows bridge was intended to sway from side to side as much as 20 feet. But even during construction, the deck also moved up and down in a rippling fashion. The undulations made bridge workers seasick. They began sucking lemons as an antidote and started calling the structure "Galloping Gertie."

Publicly, Eldridge supported Moisseiff's design, explaining to newspaper reporters and chambers of commerce that the bridge was intended to move and that, when it was finished, the movement would be of no structural consequence.

Privately, however, Eldridge and consulting engineers brought in by the state were concerned. They hoped the rippling movement would stop when the bridge deck was finished and securely fastened, but it did not. When the bridge opened on July 1, 1940, driving across it was like being on

GALLOPING GERTIE'S LAST DAY

The storm that downed Gertie was nothing out of the ordinary. The wind started picking up early in the morning of November 7, 1940, in typical fashion for Novembers in Washington. Winds blew northeast through the Narrows with speeds that reached a moderately heavy 35 to 46 miles per hour.

What was unusual was the way the bridge reacted. In addition to its usual swaying and buckling it added a new twist: Alternate sides of the deck rose and fell, like a conga dancer shrugging her shoulders.

Leonard Coatsworth, a copy editor at *The Tacoma News Tribune*, was on his way from Tacoma to his family's summer cabin on the peninsula that morning, along with his daughter's cocker spaniel, Tubby.

When he got to the bridge, shortly before 10 a.m., its strange movements had attracted a small crowd on the Tacoma side. Coatsworth watched the bridge for a few minutes, then drove under the approach and watched some more.

To him, the movement seemed odd but not dangerous. He paid his 10 cents at the toll booth and headed across. At first, the bridge merely rippled beneath him, with an exaggerated version of the wave action he had experienced before. But after he passed the Tacoma tower and drove onto the main span, the movement suddenly changed. Without warning the pavement tilted 45 degrees to the left, sending his car sliding across the bridge and into the curb on the opposite side of the road. Then, just as quickly, it tilted the other way, sending his car sliding back to the right.

Panicked, Coatsworth decided to leave his car and run back to the Tacoma side. He opened the car door and tumbled out. Immediately, the bridge

▲ Opening day, July 1, 1940. Peninsula boosters combined economic hopes with a plug for Democratic President Franklin D. Roosevelt. The sign on the car promised too much of FDR. It said, "He gave us the bridge. He will protect it."

▶ Former *News Tribune* newsman Howard Clifford, at the age of 94, holds a photo of a man running off the collapsing bridge. Clifford thinks he is the man, but the historical record is unclear.

an amusement ride. Cars in the line ahead would vanish and then reappear as they rose and fell with the waves of the bridge deck.

Eldridge and the consultants began experimenting with various tie-down systems and damping mechanisms to rein in the galloping deck. In wind tunnel tests on a bridge model at the University of Washington, Eldridge and engineering professor Frederick "Burt" Farquharson found what they thought was a solution. Cutting holes in the side girders allowed the wind to pass through the structure and stopped the erratic motion. Eldridge made plans to start cutting holes in the real bridge, but by then it was too late.

threw him onto his face and slid him across the pavement. He crawled and stumbled back to the tollbooths, now closed. His knees were bleeding, the toes of his shoes worn through from crawling. His main concern was Tubby, still trapped in the car.

But like a good newsman, he phoned his paper's city desk.

Howard Clifford was in *The Tacoma News Tribune's* newsroom when Coatsworth called. Clifford was a sort of jack-of-all trades at the paper. He did whatever needed doing, writing an occasional sports story, laying out pages, putting together the women's section. If the paper's head photographer was busy, Clifford took pictures.

When they got Coatsworth's call, the paper's editors sent reporter Bert Brintnall to the bridge and told Clifford to join him with the paper's spare camera, a big twin-lens Graflex.

More than a half-century later, when Clifford was 91, he talked about the day in the front room of his house in Normandy Park, near Seattle-Tacoma International Airport.

In addition to working 20 years in the newspaper business, Clifford had served in the U.S. Marine Corps, worked as a commercial airplane pilot, a sports commentator, a ski instructor and a race car driver. He wrote eight books, most of them about Alaska history.

Even with all those experiences, Clifford said he remembered the day Galloping Gertie fell as if it were yesterday.

"They couldn't get a hold of the regular photographer," he said. "I was told to grab the secondary camera and go out there and see what was what."

He remembered his editors told him to "absolutely run no risks."

Clifford and Brintnall raced to the Narrows and got there in time to see the strange twisting motions that Coatsworth had reported. Then, Clifford remembered, there was a weird lull. The wind let up a bit and the bridge calmed down. He decided to take the opportunity to rescue Tubby. Coatsworth was a good friend, he said, and he considered Tubby a friend, too.

"He knew me and liked me—as well as he liked the people that owned him," Clifford said. "I thought he would come to me. The thing that really spurred me on was to see if I could get the dog out. If the dog wasn't there, I probably wouldn't have gone. Or, if I didn't have a camera, I wouldn't have gone. It was a combination of the two."

Clifford walked cautiously onto the bridge deck, heading for Coatsworth's car nearly a half-mile away and pausing to snap pictures as he went. Close to the Tacoma tower, the bridge's frantic movement began again, more intense than ever.

"I got about 10 yards from the tower and I kind of looked down into the camera," Clifford said. "It almost framed it—the car, the bridge beyond it and the tower."

At 11:02 a.m., as Clifford watched through the viewfinder, 600 feet of the Gig Harbor end of the main span shook free, flipped over and fell 200 feet to the water below. Coatsworth's car, with Tubby still inside, went with it.

Clifford's thoughts went instantly from the dog's survival to his own. "When it dropped like that, I turned and started running back to shore," he said. "I didn't know if I was going to live or not, but I was going to give it a try.

"The bridge was twisting and it was bouncing because the center section was gone," Clifford said.

"It was moving faster than gravity. I would be

▲ A 40-mile-per-hour wind added an unusual twisting motion to Gertie's usual undulations. The roadway tilted back and forth almost 45 degrees from horizontal, throwing Leonard Coatsworth's car from curb to curb.

▶ At 11:02 a.m. a 600-foot section of the main span broke apart and fell into the Narrows. The rest of the main span followed shortly after, taking with it two vehicles and the only victim, a cocker spaniel named Tubby, who was too afraid to leave Coatsworth's car.

▲ Nine days after the collapse, inspectors walked along a main cable. The violently truncated roadway and twisted girders were all that remained of the main span. The bridge towers withstood the gyrations but were bent too badly to reuse.

running and the bridge would drop out from under me. Then it would come up and hit me as I was still coming down."

Clifford made it off the bridge just before the rest of the center span tore loose and fell to the Narrows. The deck he had been crawling along minutes before dropped 60 feet, then rebounded and settled into a lazy slump.

"It wasn't until later that day that I realized my trousers were torn and my knees looked like hamburger," Clifford said. "The next day I looked even worse. I was black and blue from my feet to my hips."

News about the bridge's weird undulations spread fast on the morning of the collapse. By the time it fell, hundreds of people were watching from the bluffs in Tacoma.

Eldridge was among the spectators, white-faced and appearing shellshocked. He had nothing to say to reporters about the bridge he'd help design. Nearby, Henry Foss, whose tugboats had worked all through the construction process, had tears in his eyes as he watched the bridge break apart.

Eldridge might have been at a loss for words on the day of the collapse, but the next day he had plenty to say. Pressed about why the bridge had failed, Eldridge blamed Moisseiff's redesign.

"The men who held the purse strings were the whip-crackers on the entire project," he said. "We had a tried-and-true conventional bridge design. We were told we couldn't have the necessary money without using plans furnished by an Eastern firm of engineers, chosen by the moneylenders. In order to obtain governmental money, we had to do as we were told."

Later analyses vindicated Eldridge. Had the bridge been built with the supporting trusses he put

in the original design, it would have easily stood up to the storm.

But whether or not Eldridge deserved the blame, he could not escape it. Gertie's collapse was international news. Five months later he quit his job with the state and went to work for the U.S. Navy, far away on the island of Guam in the South Pacific, looking for anonymity.

For a while he found it. The Japanese invaded Guam shortly after the December 7, 1941, attack

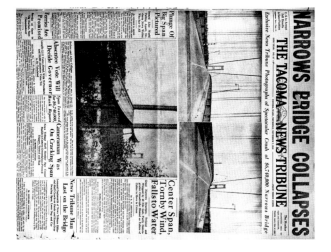

▶ The day after the bridge fell *The Tacoma News Tribune* published the pictures that Howard Clifford nearly died shooting. The next day, editorial writers blamed the failure on attempts to economize and urged a quick replacement.

▲ The state toll bridge authority's construction engineer, Clark Eldridge, left, and New York design consultant Leon Moisseiff, on the bridge six days after the collapse. Moisseiff, whose changes to Eldridge's original design were blamed for the failure, died less than three years later.

Gertie's failure took Moisseiff's reputation with it. The famous designer died of a heart attack less than three years later.

"STURDY GERTIE"

The first Narrows bridge survived only 130 days, but during that time, more people crossed it than anyone had expected.

A significant part of the traffic could be ascribed to the novelty. On sunny weekends, hundreds of people drove back and forth for the high-altitude view up and down the Narrows. On July 4 alone, 5,500 vehicles passed through the toll gates. When the wind blew, thrill seekers crossed the bridge to experience the bounce.

Tacoma boosters were quick to use the high traffic counts as evidence that the bridge should be rebuilt immediately. Members of the Tacoma Chamber of Commerce wasted no time mourning the loss. Almost before the ripples in the Narrows had subsided they began planning a successor, this time with four lanes instead of two and possibly with

on Pearl Harbor and took Eldridge captive. For nearly four years, he was in a prison camp near Kobe, Japan. But Galloping Gertie's notoriety followed him even there. One day, Eldridge later wrote, a Japanese officer recognized him, came close and said the last words he wanted to hear: "Tacoma Bridge."

As for Moisseiff, he was at a loss to explain why the bridge failed. "I'm unable to understand how it could have happened," he said. "I've built bridges all over the world for 45 years and never before has anything like this happened."

▲ Pinetree Colby, a member of a riveting crew on the deck of the 1950 bridge, was known for his consistent good humor despite his unpleasant job: bracing hot rivets while the riveter hammered them with a rivet gun.

▼ After the first bridge failed, designers took no chances with the second, which earned the nickname, "Sturdy Gertie." A truss system 33 feet deep went under the deck to increase its rigidity.

a lower deck to accommodate heavy rail traffic to the Navy shipyard at Bremerton.

But after Pearl Harbor, replacing the bridge at the Tacoma Narrows became a low national priority. The steel bridge towers, cables and what remained of the deck girders were carefully removed and recycled for the war effort.

It wasn't until after the war that everything came together to build a replacement. Engineers determined the new bridge could be a mix of old and new. The big underwater foundations were as solid as ever and could be reused without modification. The anchorages on shore also were undamaged and could be reused, though they would need to be enlarged. A portion of the Gig Harbor side span was still intact and could be incorporated into the new structure.

But more significantly, what the new bridge took from Gertie—for that matter, what every suspension bridge built after 1940 took from it—were lessons in aerodynamics. Engineers and mathematicians still debate the fine points of all that went wrong during Gertie's wild ride, but the essential failure was attributed to the flexibility of the deck and the broad, flat surface of the girders, which acted like sails. The wind hit the side of the deck structure, engineers concluded, and went into disarray as it flowed over and beneath it. Varying pressure on the top and bottom of the deck drove it up and down, inducing a condition engineers call "self excitation."

Designers of Gertie's replacement got rid of the girders and called for deep, open trusses. Based on professor Farquharson's groundbreaking work, they also called for open grates in the concrete road surface to relieve air pressure.

Construction on the new bridge began in April 1948 and took 30 months to finish. The structure was so stable in the wind a few started calling it "Sturdy Gertie," a name that, though appropriate, never really took hold.

The replacement bridge opened to traffic on October 14, 1950, and was moderately heavily used in the postwar years. The first year it was open, an average of 4,000 vehicles crossed it each day. Twenty years later, more than five times that many people were using the bridge, but still there were no congestion problems. West of the bridge, Highway 16 quickly meandered into a two-lane country road. Most days, drivers could cross the bridge from Tacoma and hang a leisurely left across the sparse eastbound traffic to get burgers and shakes at the Span Deli, a popular drive-in on the Gig Harbor approach.

But during the late 1980s and 1990s, Tacoma's long predicted "renaissance" at last moved toward reality. Economic benefits from the computer technology revolution in Seattle and King County spilled south into Tacoma and Pierce County and created jobs.

The city retained its gritty blue-collar nature during the boom, however, maintaining crime rates that stubbornly remained among the highest in the country. Most young professionals attracted by jobs in the city chose to live away from its core. The Gig Harbor area, with its forests, water views and crime-free neighborhoods, looked to many like a safe haven.

As developers shaved forests west of the Narrows to accommodate suburbs, bridge traffic soared. The lanes on the bridge were 2½ feet narrower than the 12-foot federal standard, and a row of orange plastic pickets was all that separated eastbound and westbound traffic. Many drivers found the open grates installed for air flow creepy to drive across, increasing their anxiety.

GERTIE'S LESSON

The failure of the first bridge across the Tacoma Narrows forever changed concept of aerodynamic stability in suspension bridges.

"GALLOPING GERTIE"
The deck of the 1940 bridge, now at the bottom of the Narrows, took the trend of light, graceful design too far. The deck had no stiffening trusses, and when the wind hit the flat face of its 8-foot-tall side girders, it produced violent oscillations.

43'

8'

"STURDY GERTIE"
Designers of the 1950 bridge were determined not to repeat Gertie's mistakes. The deck has a 33-foot-tall truss system for rigidity and open grates in the roadway to allow air to pass through.

60'

33'

In 1980, an average of 38,973 vehicles crossed the bridge each day; in 1990 the average was up to 66,573. By 2000, when the wheel fell off Marvin Parkko's logging truck, 88,000 cars and trucks a day were squeezing through the bridge's four narrow lanes. Long backups during the morning and evening commutes were routine.

While most agreed that a second bridge was needed, nobody could figure out how to pay the hundreds of millions it would cost. Reluctant to raise its gas tax, the state did not have enough money even for routine fixes to its existing highways, let alone the kind of money necessary to build a new Narrows bridge. The state's financing outlook deteriorated further in 1992 when voters forced the Legislature to toss out the other major source of transportation funding, the motor vehicle excise tax.

The prospect of paying for the new bridge with tolls met bitter protest in Gig Harbor and nearby communities on the west side of the Narrows. If a bridge were built, they argued, the cost should be paid for by everybody in the state, like all other public transportation. Tolls would unfairly burden those who happened to live next to it. The suggested $3-per-day, round-trip fare would cost regular commuters hundreds of dollars a year, they protested, and would suck the economic life out of the Gig Harbor economy.

The communes at Home and Burley were long gone, but the old separatist sensibilities were alive and well. A fair number of west-siders did not want a new bridge at all, cherishing their isolation and fearing a second span would bring a stream of big-box chain stores and commercial development. The

"We feel like we're dealing with King George and we are the colonists."

—Gig Harbor resident

daily struggle on the bridge, they believed, was a price worth paying to prevent urban sprawl.

A strong and loud group of anti-bridge residents sprang up on the peninsula, holding bake sales and selling T-shirts to finance their cause. But they were outspent nine-to-one by a coalition of contractors, real estate agents, builders and labor union members who backed a "Yes on Tacoma Narrows Bridge" campaign.

Joe Gotchy, an ironworker who had worked on both Narrows bridges and wrote a book about the experience, was among the supporters of a new bridge. Then a leathery 89 years old, Gotchy carried petitions around to stores in the Gig Harbor area, urging the state to begin work quickly on another Narrows bridge.

"I'd like to kick them in the butt and get them started," Gotchy said.

To help resolve the conflict over tolls, the state called an advisory vote. In an election that included parts of seven counties near the bridge and more than 90 percent of the people who used the bridge at least occasionally, 53 percent said yes to tolls. In the Gig Harbor and Key Peninsula areas, where most regular bridge commuters lived, 83 percent voted no. After they lost, deep resentment settled in.

"The feeling in this community overwhelmingly is one of betrayal," a Gig Harbor woman named Elena Lindquist said at one of dozens of public gatherings on the issue. "We feel like we're dealing with King George and we are the colonists. We need to have a Gig Harbor Tea Party."

The protests faded with defeats in court. And, after Marvin Parkko's logging truck accident, oppo-

► By 2000, most people agreed another bridge was a good idea. Paying for it was a different matter. Peninsula residents strenuously objected to tolls because most of the burden of paying for the $849 million bridge project would fall on them.

▲ Heavy side winds and rain often made crossing the 1950 bridge a harrowing experience. Open steel grates on the road surface, ordered by designers to dampen aerodynamic forces, increased some drivers' trepidation.

▲ Governor Gary Locke and Girl Scouts toss dirt at a ceremonial groundbreaking at the new bridge's Tacoma anchorage. In the background, protesters object to plans to make one of the lanes on the new bridge open only to vehicles carrying more than one person.

nents receded into sulky silence. The new bridge was going forward whether they liked it or not. The state signed a $615 million deal with the Bechtel Corporation and Peter Kiewit Sons Inc. to design and build a new bridge and fix up the old one. The total cost of the project, $849 million, would be financed by government bonds repaid by tolls.

At the groundbreaking ceremony on Saturday morning, October 5, 2002, about 200 people gathered at the Tacoma end of the existing bridge to celebrate. The ceremony had all the right ingredients—a brass band, chrome-plated shovels, a fluttering American flag and dozens of politicians beaming for cameras. About 100 spectators sat on folding chairs under a blue awning set up for the occasion, and another 100 or so stood on the perimeter, munching free sweet rolls and drinking coffee from paper cups.

Congressman Norm Dicks, whose district covered both sides of the Narrows, said he was pleased that the contentious planning of the project was over and that construction could begin.

"The bottom line is, it's done and we're moving ahead," Dicks said.

Governor Gary Locke called the groundbreaking "a great moment for the state of Washington."

After the speeches, politicians and dignitaries moved to a small roped-off area where a truckload of loamy topsoil and 30 shiny shovels had been deposited. There, they took turns digging and posing for photos.

The spot where they dug soon would be excavated to deposit 23,000 cubic yards of concrete that would anchor the Tacoma end of the new bridge.

But even as construction moved forward, opposition continued. A handful of protesters stood quietly on the sidelines holding signs objecting to bridge tolls.

"The fight isn't over yet," one of them insisted.

▲ **Skeptical peninsula residents listen to plans for a new bridge. Lawsuits and disputes over financing held up construction for years.**

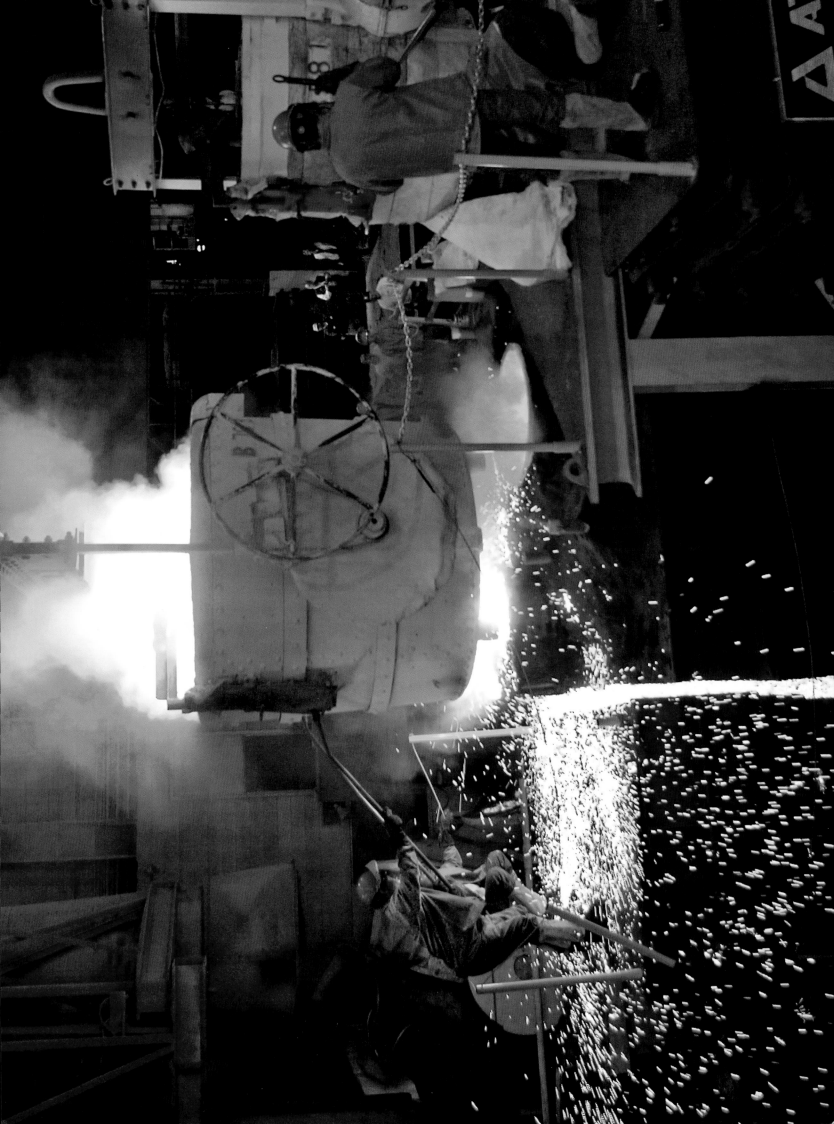

FORGING THE TEAM

Led by an enthusiastic Venezuelan engineer, Tacoma Narrows Constructors gathers top bridge-building talent from down the street and around the world.

When Manuel Rondón arrived in Gig Harbor, he sometimes felt like an invader living behind enemy lines. He had been chosen to run the Narrows bridge project, and for him it was the opportunity of a lifetime. The new bridge would be the largest suspension bridge built in the United States in 40 years, and it meant a chance to manage an international dream team of construction industry talent.

But in 2000, when he moved with his wife and three children to a house in Gig Harbor, he found himself surrounded by people who thought the bridge was nothing but an expensive mistake.

At the barbershop, Rondón, bursting with pride, had to sit and listen to people complain about tolls and about the traffic congestion that five years of construction was going to cause.

The reach of the anti-bridge feeling became clear one day when his 9-year-old son, Gabriel, came home from fourth grade at Artondale Elementary School and asked, "Is it OK to say you work with the bridge?"

Rondón, indignant, told him yes. "Tell them we don't have anything to do with how the bridge is going to be paid for," he said. "We're just building it."

Rondón was born and raised in Caracas, Venezuela. At 47, he had a long history of managing complex construction projects, mostly in South America's petrochemical and aluminum industries.

▲ **One of several Puget Sound-area companies that won contracts to help build the new bridge was Tacoma's Atlas Castings & Technology foundry. Here, foundry workers pour molten steel into forms for the splay saddles that would cradle the bridge's suspension cables at each of the two anchorages.**

His experience with bridges was limited, but impressive. The construction giant Bechtel, one of the two main partners in building the Narrows bridge, had found him in Portugal, where he had just finished managing an extensive retrofit of Europe's longest suspension span, the 3,323-foot Tagus River bridge in Lisbon.

That job occasionally involved dealing with irate people, too. He was to add a lower deck on the Tagus bridge for freight trains and commuter trains, plus add a sixth lane of traffic to the top level, all while 140,000 vehicles a day continued to use the bridge. Miscalculations meant traffic jams and angry drivers.

The anti-bridge fervor in Gig Harbor bothered Rondón, but it didn't surprise him. He regarded the situation in what, for him, was a characteristic manner: He saw it with an engineer's unflappable practicality, combined with a personal tendency to take the long view.

The tolls, he could do nothing about. As for congestion, he said he intended do everything he could to keep traffic flowing smoothly, but the fact was, urban construction always caused disruption.

"It's going to be a big inconvenience for people," he said. "I would be upset, too. Building a bridge is like having a baby. No matter what you do, it requires a certain amount of time and involves a certain amount of pain."

Mostly, Rondón wanted to bring people around to his point of view—that they were fortunate to be living next to one of the most fascinating construction projects in the world. They all would have front-row seats for a once-in-a-lifetime production.

Rondón was not an especially large man, but his self-certainty demanded a lot of space. At restaurant tables filled with engineers and construction superintendents, waiters never needed to wonder who

should get the check. Wherever he was present, he presided.

His English, though grammatically faultless, was richly rounded with Caracas Spanish. The accent, in concert with his gray-streaked beard, worked in his favor, giving him exotic status. People didn't immediately know what to make of him.

When Rondón wanted to, he could quickly put people at ease. He could be charming. He loved conversation, classical music, good wine and food. As soon as he arrived in Washington he started seeking out Seattle Symphony tickets and scouting out gourmet restaurants. At the Starbucks where he went for coffee, he knew the barristas by name.

But he also had an unpredictable quality, like a bear that might suddenly turn on its trainer. And he could be obsessive about detail. The engineers who

▶ Manuel Rondón, the Venezuelan engineer chosen by the Bechtel Corporation to manage construction of the bridge, arrived with a reputation as a tough, no-nonsense negotiator. He also turned out to have a sense of humor. Here, he enjoys a laugh with his project engineer, Scott Steingraber, left, and construction manager Pat Soderberg.

▲ Rondón leads an early morning safety meeting for workers at the Tacoma Narrows Constructors' field office near Gig Harbor.

ral politician and well suited to the task. MacDonald had honed his public pacification skills as head of the Massachusetts Water Resource Authority, where he successfully led a controversial $3.8 billion cleanup of Boston's polluted harbor, financed by raising people's water and sewer bills.

MacDonald assembled a strong public relations team to win converts in Gig Harbor. Their efforts included a blizzard of informational brochures and scores of public meetings where people could talk to engineers and examine models of the bridge. They set up a sophisticated and thorough Web site with remote camera feeds so residents could monitor traffic and watch bridge construction from their home computers.

Linea Laird, the manager of the state team of engineers and accountants selected to oversee Rondón's work, was less politically inclined than MacDonald, but she did her best to establish herself as someone willing to listen. She acceded with little fuss to many local demands, agreeing to cut fewer trees and build noise barriers to protect neighborhoods. When people complained about the 13-lane bridge-approach corridor that would slash through Gig Harbor's forests, she signed off on a $1.46 million change order to add more landscaping.

Rondón, meanwhile, focused on organizing the worldwide web of talent it would take to build the bridge—designers, engineers, steel suppliers and fabricators, tugboat operators, test labs and labor unions.

Bechtel and Kiewit were the main partners in the team, having joined in a temporary partnership named Tacoma Narrows Constructors. The network of subcontractors they assembled came from as close as five miles from the bridge to as far as 5,000 miles away, ranging from Japan and South Korea to the

joined him in Gig Harbor told stories of being fined for being a minute late to a meeting or for letting a cell phone ring. In the interest of safety, they were curtly told not to tip backward on conference room chairs.

Rondón's reputation as an adamant, unyielding negotiator preceded him. The German company, DSD Dillinger Stahlbau GmbH, which headed the Tagus River bridge retrofit, hired Rondón after sitting across the negotiating table with him on a previous job. They told Rondón, "We want you to do for us what you were doing for them."

Tom Draeger, a senior vice president at Bechtel's headquarters in San Francisco and one of the executives who selected Rondón for the Narrows project, put it this way: "What they say here is, 'I wouldn't want to piss him off too much.'"

Rondón was trained as an electrical engineer but decided early in his career that he preferred running things. He learned, as he was fond of saying, "It is better to be the head of the mouse than the tail of the lion."

While Rondón recognized the need for good community relations at the Narrows, that was not his personal forte or even close to his top priority. He was asked innumerable times to make presentations to this or that civic group. For the most part, he dodged the invitations. At community meetings called to explain the construction process, he usually sat quietly in the audience, wearing a woodsy flannel shirt to blend in, while his engineers did the talking. When Governor Gary Locke and his staff wanted a briefing on the bridge project, Rondón sent one of his managers instead of going himself.

The brunt of the public relations work fell to the Washington State Department of Transportation. The agency secretary, Doug MacDonald, was a natu-

▶ **Ironworkers wrestle rebar into position for bridge caissons. American manufacturers and laborers initially were upset because the Bechtel/Kiewit partnership called Tacoma Narrows Constructors was buying most of the steel for the new bridge from Japan and that fabrication of major bridge components would take place in South Korea.**

However, the promise of hundreds of union jobs for ironworkers, carpenters and laborers—jobs that would last five years and include plenty of lucrative overtime—alleviated their concerns.

United Kingdom and Canada. In the interest of team unity, the fact that the project would be managed by two of the world's biggest construction firms was all but buried. All visual references to Bechtel and most to Kiewit were banished from the job site and replaced with the TNC logo.

The high profile of the Narrows project made team building easier for Rondón. Suspension bridges hold magic for many people, but particularly for those involved in large-scale construction. In their world, suspension bridges constitute a separate kingdom, a place where craft comes close to fine art.

Like spider webs, suspension bridges illustrate engineering principles in a particularly pure form. They are at once extraordinarily complex and absolutely simple. The harplike array of suspender wires looks as if it might make beautiful music if strummed. The easy drape of the cables between tower tops draws a natural, evocative line, satisfying and somehow inspiring.

"They just look right," said Dave Climie, an easygoing Scotsman who eagerly accepted an invitation to join Rondón's team.

Suspension bridges are not only beautiful, Climie said, but they have what he called "the whoa factor"—that is, "They make people stop in their tracks and say, 'Whoa! How did they do that?'"

Climie, who would be in charge of all the intricate cable work and assembling the finished bridge deck, was a member of an elite international group of suspension bridge experts who traveled from continent to continent, drawn to bridge projects like guest artists at local theater productions.

Big suspension bridges were built so rarely that the talent pool was relatively small. Many of the top engineers knew one another. They had their own

conventions, their own vocabulary and even their own set of specialized machinery, some of which traveled around the world with them.

In 2000, when Rondón was gathering forces for the Narrows project, Climie, then 39, was at the top of his game. During the previous 10 years he had directed the cable work on three of the five largest suspension bridges in the world, including the 5,328-foot Great Belt Bridge in Denmark.

A Bechtel headhunter found him in China, where he was working on the Jiangyin Bridge—with the world's fourth-longest main span—on the Yangtze River.

The Narrows project appealed to Climie for several reasons. Size was part of it. The main span—the length of deck suspended between towers—would be 2,800 feet. That was 1,743 feet shorter than the Jiangyin, but still respectable. And, Climie said, he liked the idea that the new bridge would be so visible. It was to be built just 60 feet from the existing bridge, meaning his crews would be spinning the web of cables in full view of thousands of commuters each day. "It's like a performance," he said.

Climie studied civil and structural engineering in college, but he said that is not where one learns how to build suspension bridges. In his field of construction, perhaps more than any other, knowledge gained by experience is passed down from generation to generation.

"There's no substitution for experience when you're out on the bridge at night, things are not going right, and you're deciding, what are we going to do," Climie said. "There's nothing in the books. It helps to have been in a similar situation before."

The people who design suspension bridges are as tightly tied to historical precedent as those who build them. Advances in design over the past 200 years edged forward bridge by bridge, sometimes by colossal failure—Galloping Gertie being the prime example. Famous bridges are aligned with individual designers and their personal theories, and current designers place themselves in that progression of ideas.

Thomas Spoth, then 38, was chosen to head the Narrows bridge design team. The Parsons Transportation Group engineer tracked his professional connections all the way back to John A. Roebling, the original champion of the monumental Brooklyn Bridge, built over New York's East River in the 19th century.

Spoth, who would manage about 60 designers and engineers working on the Narrows project in offices in Manhattan, San Francisco and Gig Harbor, said he had been fascinated by bridges since he was a child.

When he was still in grade school, he built a bridge across a creek in his parents' backyard in suburban Washington, D.C. By the time he was in high school, he knew he wanted to build bridges professionally, and he never strayed from the career path that took him there.

"I knew I wanted to work on bridges," he said, "and I knew I wanted to work on suspension bridges."

After taking a degree in structural engineering, Spoth landed a job at Steinman Boynton Gronquist & Birdsall, the New York firm founded by the legendary bridge engineer David Steinman, whose projects included the Mackinac Bridge, connecting Michigan's upper and lower peninsulas. Once on the staff, Spoth was thrilled to meet Blair "Mr. Bridges" Birdsall and Jack Nixon, two top engineers who had once worked for the design firm founded by Roebling.

▲ Dave Climie, an enthusiastic and talented Scot, joined the Narrows team as head of superstructure construction. Climie was regarded as one of world's foremost suspension bridge engineers, having directed cable work on three of the five largest suspension bridges in the world.

He liked building suspension bridges, he said, in part because of "the whoa factor." They make people stop in their tracks, he said, and exclaim, "Whoa! How did they do that?"

construction, had worked on all of Puget Sound's floating bridges, but never on a suspension bridge. He was attracted by the difficulty of building the bridge's underwater tower foundations, or caissons. They would be as big as 20-story office towers and extend 200 feet below the water surface into a channel notorious for its rapid tidal currents.

Unlike Rondón's other top builders, Sherman had never been to engineering school. He learned his trade through experience, beginning with a carpenter's apprenticeship. He also had less far to travel than any other top team members. He lived just 20 miles from the Narrows on Vashon Island, where he could trace his lineage back five generations.

Sherman happily accepted a job as the superintendent of caisson construction.

"These are the largest caissons built anywhere in the last 40 years," he said. "This is a once-in-a-lifetime opportunity for somebody in the construction business. It's a signature project. Five to 10 years from now, I want to turn on the TV and see it on 'The Learning Channel.'"

When Spoth was assigned to the Narrows job, it gave him a satisfying sense of historical continuity: Birdsall and Nixon had worked on the design of the 1950 Narrows bridge. There was a more recent connection, too, one with Rondón. Parsons, Spoth's firm, had designed and engineered the Tagus bridge retrofit.

"Those of us working on bridges now picked up this torch and that knowledge from these older construction trades—some whose ancestors had worked on the old bridge—Gertie's notoriety made them eager to take part. They competed for jobs on the new bridge, many seeing it as a chance for immortality. Someday their grandchildren would look at the bridge, they believed, and say proudly, 'My grandpa helped build this.'"

The historic failure of Galloping Gertie also added to its appeal. For engineers especially, the Narrows bridge was famous. Its failure was iconic; the fact that it happened to be captured on film made it a classic. Engineering students all over the world watched the old black-and-white film in their classes on the effect of aerodynamics on suspended structures.

For Tacoma-area ironworkers and others in the construction trades—some whose ancestors had worked on the old bridge—Gertie's notoriety made them eager to take part. They competed for jobs on the new bridge, many seeing it as a chance for immortality. Someday their grandchildren would look at the bridge, they believed, and say proudly, "My grandpa helped build this."

The technical challenges presented by the Narrows project also helped attract top people to Rondón's team. Tom Sherman, a 57-year-old native Northwesterner and an authority on marine

▲ The Foss Maritime Company, which handled tugboat and hauling tasks for both previous bridges, won similar contracts on the new bridge.

▶ Washington trade workers, steeped in the lore of Galloping Gertie, competed for jobs on the new bridge and a place in history. Adrianne Moore thrived in one of the most physically demanding jobs, handling the concrete that went into the caissons and towers.

Adrianne Moore remembered seeing the old Galloping Gertie film in high school and was thrilled when she landed a job as a journeyman laborer on the new bridge. Part of what attracted her was the prospect of five years of steady work just a 10-minute drive from her house in west Tacoma. But in addition to that, she said, it was the historic stature of the old bridge.

Moore's job would be the hardest kind of physical labor, helping place the thousands of yards of concrete that would go into the new bridge's foundations and towers. The prospect didn't faze her. At 34, she'd been in the construction industry for 12 years and was proud of the physical stamina that qualified her as a member of the Laborers Union. "We're pack mules," she said proudly.

She campaigned hard for her bridge job, using what she called "the doughnut plan." Early each morning for several days, Moore, dressed in her boots and hard hat and ready to work, took doughnuts to the job site. She handed them out as she pleaded her case to people already on the crew. "I had to beg and beg and beg," she said.

One day when the concrete superintendent was within earshot, she yelled, "Hey, buddy. If you're looking for the best hand in town, here I am." The superintendent hired her. And she eagerly took her place in history.

"A hundred years from now people will open *The News Tribune* and see pictures of me sitting on top of the bridge," she said, "just like all those old pictures."

The physical risk implicit in bridge building also had its appeal. Building suspension bridges is dangerous work. It was nothing like it had been during America's great bridge-building decades of the late 19th and early 20th centuries, when the rule of

thumb was one man killed for every million dollars spent. But construction that took place hundreds of feet over rapidly moving water still carried deadly risks. Machismo was clearly a motive for old-time bridge-builders who posed nonchalantly on narrow beams with hundreds of feet of empty space below.

Accurate records of deaths weren't kept during the building of the Brooklyn Bridge, but at least 20 men died on that job and possibly as many as 30. Eleven people died building the Golden Gate Bridge. The death toll in San Francisco would have been higher had superintendents not strung a net below the cables, like the ones for trapeze artists in circuses. Nineteen men fell into the net instead of San Francisco Bay, qualifying them as members of what became known as the "Halfway to Hell Club."

One man died building Galloping Gertie. A carpenter named Fred Wild fell 12 feet from scaffolding just five days before the bridge's grand opening July 1, 1940. Three men died building the 1950 bridge: Robert Drake was hit by a falling boom in a crane accident on shore, and ironworkers Stuart Gale and Whitey Davis fell into the Narrows from nearly 200 feet as they were helping with the midair assembly of the deck. Their heavy tool belts carried them quickly to the bottom.

Rondón was determined that the death toll on the new bridge would be zero.

"There is nothing more difficult than going to somebody's family and saying he or she is not coming back tonight," he said. "What is it worth at the end of the day if you sacrifice a human life?"

On the Tagus River bridge job in Portugal, two workers died, despite an aggressive safety campaign. As project manager, it was Rondón's job to bring men's families into the office and explain to them what had happened.

▲ Ironworkers in the old days of suspension bridge building celebrated a culture of bravado that came from working in dangerous, exposed positions, often without safety equipment. That was not to be the case on the new bridge. TNC approached safety almost obsessively, vowing to do everything possible to make sure no one was killed or seriously injured during construction.

"It is beyond words the emotions that permeate a meeting like this," he said.

MILLIONS FOR LOCAL LABOR

The public relations problems Rondón faced in Washington were compounded when it became known that Tacoma Narrows Constructors did not intend to use American steel for the project.

All the concrete reinforcing bars, all the steel for the mile-long deck and the 5,500 tons of wire in the main cables were to be bought from the Japanese giant, Nippon Steel. In addition, the deck fabrication, a two-year cutting and welding project, would be handled not in the United States but by Samsung Heavy Industries in South Korea.

American manufacturers protested, but because no federal funding would be going into the project, TNC was exempt from the "Buy America" requirements that would have required iron and steel products be manufactured in the United States.

Without government intervention, American steel companies—what was left of them—could not begin to match the bids of Asian firms. Buying American would have added $30 million to $40 million to the cost. The Asian companies not only had cheaper labor but they also had industrial facilities already geared up to produce the massive bridge orders.

Asia also had more suspension bridge expertise. Of the 10 longest-span suspension bridges completed in the world since 1996, eight were in either China or Japan. They included the world's longest, Japan's $5 billion Akashi-Kaikyo Bridge with a main span of 6,529 feet, more than twice the length of the one required at the Tacoma Narrows.

TNC's deals with Asian companies also angered American labor unions. The deck construction alone would provide jobs for hundreds of Korean workers for two years.

Unions didn't like what they saw as exporting jobs, but they were mollified by the number of jobs the bridge would create locally. Over the five years of construction, the project would provide hundreds of thousands of hours of union jobs, with overtime and benefits. TNC signed a project labor agreement requiring all directly hired workers on the construction site be union members.

And, while the biggest contracts went to Asian subcontractors, the bridge project would spread millions of dollars to Washington state companies as well, ranging from a three-person model-making shop in Tacoma's Nalley Valley to Seattle's Todd Pacific Shipyards.

Todd, on Harbor Island at the mouth of the Duwamish River in Seattle, was selected to build the bridge's "cutting edges," the steel caisson bottoms. Its $5.4 million contract was only one-fifth the size of the one awarded to Samsung, but it was nevertheless a welcome boost for Todd, still recovering from a bankruptcy reorganization.

Tugboats in the Foss Maritime Company's fleet would handle all the heavy towing work for the new bridge, just as they had for Galloping Gertie nearly a half-century before.

In Tacoma, the Atlas Castings & Technology foundry won the contract to forge the bridge's "saddles," the massive steel forms, as big as small trucks, that would cradle the main cables at the tower tops and anchorages. To handle the job, Atlas added 15 employees to its work force, including molders, core makers, welders and grinders.

Jesse Engineering Company, a steel fabrication firm near the Port of Tacoma, was hired to make other specialized equipment, including the big

spinning wheels that would carry main cable wire back and forth across the Narrows. And Amaya Electric Company, based just south of Tacoma in Lakewood, became the primary electrical contractor on the job.

Strictly speaking, the biggest local contract was not local. Glacier Northwest Incorporated won the contract to provide the 190,000 cubic yards of

► **Even with stringent safety safeguards, work on suspension bridges requires exposure not appealing to everyone. Here, electricians installing work lights negotiate the temporary construction catwalks hundreds of feet above the Narrows.**

concrete that would go into the bridge's anchorages, caissons, towers and highway approaches. It was enough concrete to lay down a highway from downtown Tacoma to Seattle, 30 feet wide and a foot deep. However, while Glacier Northwest was based in Seattle, its owners were in Tokyo: the mammoth Taiheiyo Cement Corporation.

TAKING THE LONG VIEW

In July 2003, after more than two years of planning, hiring and preparation, construction was at last about to begin on the water in the Narrows. One sunny afternoon, Rondón took a break from the TNC offices and drove to a small waterfront park just south of the bridge site on the Gig Harbor side.

The spot, which most locals called Doc Weathers Park in memory of a former owner, had a sandy beach accessible by a dirt trail through blackberry bushes. For Rondón, the attraction was the view. From the beach there was a sweeping shot of the old bridge, the mile-long arc of its cables visible all the way from one anchorage to the other.

Directly in front of it was the construction site for the new bridge—nothing but open water at that point, but soon to be bristling with construction cranes and crowded with tugboats and barges.

The tide was slack in the Narrows that afternoon and the beach was almost deserted. A woman, sunbathing, lay face down on a towel. Nearby, two dogs chased each other in and out of the water. On the bridge, lines of tiny cars and trucks moved in unison, like trains pulled by faraway engines.

The setting was perfect for Rondón to philosophize, and he did not disappoint. For him, the bridge project was not just a bridge, he said, it was an experience, soon to be a memory.

"Moments like this are what we will take from this project," he said. "When it is all over we will look back and remember what it was like before and see what we have done."

One of his great satisfactions in Lisbon, Rondón said, after the Tagus bridge was finished and the construction material cleared away, was going back to a favorite restaurant on the river, where he often ate lunch during the project.

He would take a table by a window where he could see the bridge soaring above him. It gave him indescribable pleasure, he said, sitting there and knowing he had some part in building it.

"I would think, 'My God, when we were first starting . . . now look at it! Who would know that I had something to do with that bridge?' That is my satisfaction—just to know it's there."

Rondón said he intended to conduct himself on the project as if he were already looking backward, pleased with how things had gone. He wanted to create a valuable, enjoyable experience for everyone involved, so all would be able to look back with satisfaction and pride.

"For most of us, this is the first time we will have done anything like this," he said. "For many of us, it will be the last as well. Think about it. When is another opportunity like this going to come up?

"We are moving into a different world now—into what is going to be. There is history in the making here."

▲ **The new bridge would forever change the familiar profile of the single, delicate span across the Narrows. "We are moving into a different world now," Manuel Rondón said as construction began.**

THE WITCHES BEHIND THE DOOR

After 40 years of marine construction, Tom Sherman thought he had seen it all. Sinking skyscraper-sized caissons in the Narrows' rapid currents made him think again.

Tom Sherman didn't tell many people, but he had decided that building the caissons for the new Tacoma Narrows bridge would be his last big project.

He was about to turn 60 and had spent 40 years in marine construction, working with tugboats, floating cranes and pile drivers in just about every corner of Puget Sound. He had built floating bridges, marinas, ferry terminals—even an aircraft carrier pier.

But in terms of size and complexity, the bridge caissons—the foundations for the new bridge towers and the key to the entire structure's stability—were a step beyond any of those. They would be the biggest, most complex and, with any luck, the most permanent structures he had ever built—a perfect capstone to his career.

As a personal statement, the bridge project could not have been better located: The Narrows lay at the heart of Sherman's ancestral geography. His great grandparents, Salmon and Eliza Sherman, were among the first three families to homestead on Vashon Island back in 1877, riding the Narrows currents north from Steilacoom.

Sherman grew up on the island in the 1940s and 1950s, surrounded by three generations of aunts and uncles and cousins. He went to work for a Vashon poultry farmer at 13, taking

▲ **Like spiders on webs, ironworkers weave dense screens of reinforcing steel that will be encased in the walls of the caisson on the Tacoma side of the bridge.**

▲ Tom Sherman, right, explains caisson construction to Governor Gary Locke. Sherman brought 40 years of experience in marine construction to his job as caisson construction manager.

care of chickens and, when he was old enough to drive, ferrying eggs to restaurants and grocery stores in Seattle. After high school he went directly into construction, never giving a serious thought to college.

"It wasn't part of our family tradition," he said. "And I didn't have the right kind of grades."

Instead, Sherman went to work as a carpenter's apprentice at the Seattle-based marine construction firm General Construction, where he stayed his entire career. By the time Kiewit bought General Construction in 2001, Sherman had worked his way up to senior vice president. Sherman taking the Narrows job was among the terms of the sale. Sherman wanted the job, and Kiewit was happy to oblige him.

Sherman, a plain-spoken, practical man, looked young for his age. He had no sign of a paunch, and it took a close look to see that his hair was gray and not still blond. His eyes were bright blue, set in the kind of northern European skin that does not like sun.

It was unusual for someone without an engineering degree to be put in charge of building the caissons, generally agreed to be the most challenging and dangerous part of the bridge project. But Tacoma Narrows Constructors had plenty of engineers. The skills Sherman brought were practical ones, and rarer. He knew the machinery, he knew the workers and he knew Puget Sound weather and tides.

Sherman began when the project was still in the head-scratching phase, and TNC's design team gave him plenty of freedom to contribute. "We'd come up with the what-ifs," he said, "and the engineers would put the numbers on it."

Building the caissons would take all of Sherman's skills. Structurally, they were simple enough: two rectangular concrete columns, 20 stories tall, with caps on top about the size of basketball courts. The caissons would extend down through the water about 150 feet, then another 50 or 60 feet into the sea floor. Inside, they would be mostly hollow. Imagine high-rise buildings made up of nothing but elevator shafts.

The hard part was that the caissons would need to be built as they floated a quarter-mile from shore. The idea was to build them from the top down. As workers set forms and poured concrete at the surface, the entire structure would settle deeper into the

▶ Powerful, complicated currents in the Narrows would make it difficult to keep the floating caissons in position. Water rushing past the old bridge piers churned the water into swirling vortexes at the construction site.

THE NARROWS JET

The Tacoma Narrows is a natural funnel that restricts, and therefore accelerates, tidal water that flows between the main body of Puget Sound and its southern basin.

During incoming tides, water from the main basin races south through the Narrows, reaching speeds as high as 10 miles per hour.

The Narrows is 600 feet shallower than the main basin, so water is restricted vertically as well as horizontally.

Seattle

Gig Harbor

1950 bridge

Tacoma

New bridge

THE NARROWS

Lacey

The entire tidal exchange from the 173-square-mile southern basin must squeeze north through the Narrows when the tide goes out.

---

water. By the time the big boxes neared the bottom they would weigh 33,000 tons each. At that point they would have to be settled gently down onto the sea floor, within inches of a target specified by the designers, then driven into the sand and gravel.

The construction technique was unusual but by no means unprecedented. After a long experimental period—marked by sometimes tragic mistakes—it had become the standard engineering sequence for caissons. The original efforts cost many lives. In the 1870s, the first big bridge caissons built in America, for the Eads Bridge in St. Louis and for New York's Brooklyn Bridge, were lowered by men with shovels, working in pressurized compartments beneath the giant structures. When they got beyond 70 feet, dozens of workers were paralyzed or killed by the then mysterious "caisson disease," now commonly called the bends and known to be caused by rapid reduction of atmospheric pressure.

What was unusual about the Narrows job was that the caissons would be so large—they were designed to withstand 9.1-magnitude earthquakes and support a second bridge deck if necessary—and they would need to be built in extreme tidal conditions.

"Caissons are not built in this kind of environment very often," Sherman said. "In most places, crews are building caissons in lakes or rivers where water currents flow in one direction and there are only small, if any, tide fluctuations. In the Narrows, we have tides that change water levels up to 18 feet and have strong currents that flow in both directions."

As the caissons grew downward, the force of the tidal currents pushing back and forth on their wide, flat surfaces would make them increasingly hard to control. They were to be built just 60 feet from the

old bridge. If one of the caissons broke loose, it could tip over, fill with water and sink to the bottom.

Or, worse, it could smash into the old bridge and send it down alongside Galloping Gertie. Sherman didn't mind the idea of going down in history with the Narrows project, but that wasn't the way he wanted to do it.

The key to controlling the caissons, Sherman said, was anchoring, the same as it was with all construction projects in tidal water.

"Basically," he said, "what it comes down to is the ability to hold the item that you're building, plus the equipment you're building it with, while you're in construction."

Tidal currents are extreme at the Narrows because it is a natural funnel. Puget Sound has a pinched waist there, narrowing to a distance of about a mile from an average of six to seven miles farther north and south. When the tides surge in or out, the Narrows speeds the water up, like a jet nozzle on a hose. Currents sometimes move at more than 10 miles per hour, shooting boats through the channel like fourth-graders on a Wild Waves ride.

In 1939, crews building the caissons for Galloping Gertie—the ones still in use on the 1950 bridge—had a hard time keeping the currents from carrying their construction barges away. Once anchored, the 100-foot barges created serious navigation hazards. Midway through construction, an incoming tide pulled a fishing boat against one of the barges. The current sucked the boat and two fishermen under the barge and shot them out the other end. They survived only because quick-thinking construction workers slid out a life raft for them.

The funneling effect at the Narrows is intensified because the channel is situated on top of an under-

SKYSCRAPERS IN THE SEA
The caissons were like high-rise buildings constructed upside down. Rapid tidal currents made the job the most technically difficult part of the bridge project.

PLACING THE ANCHORS

During construction, each of the caissons was tethered to 32 anchors, carefully positioned to maximize protection from currents. The anchors were arranged in two concentric circles, one 300 feet and the other 600 feet from the caissons.

GIG HARBOR

NEW BRIDGE

CURRENT BRIDGE

N►

TACOMA

Old caisson

New caisson

- • Outer anchors
- • Inner anchors
- — Chains

Construction of each caisson began with a "cutting edge" that eventually became the base of the structure. The cutting edge contained 15 closed chambers that were cut open to allow dredging when the caisson hit bottom.

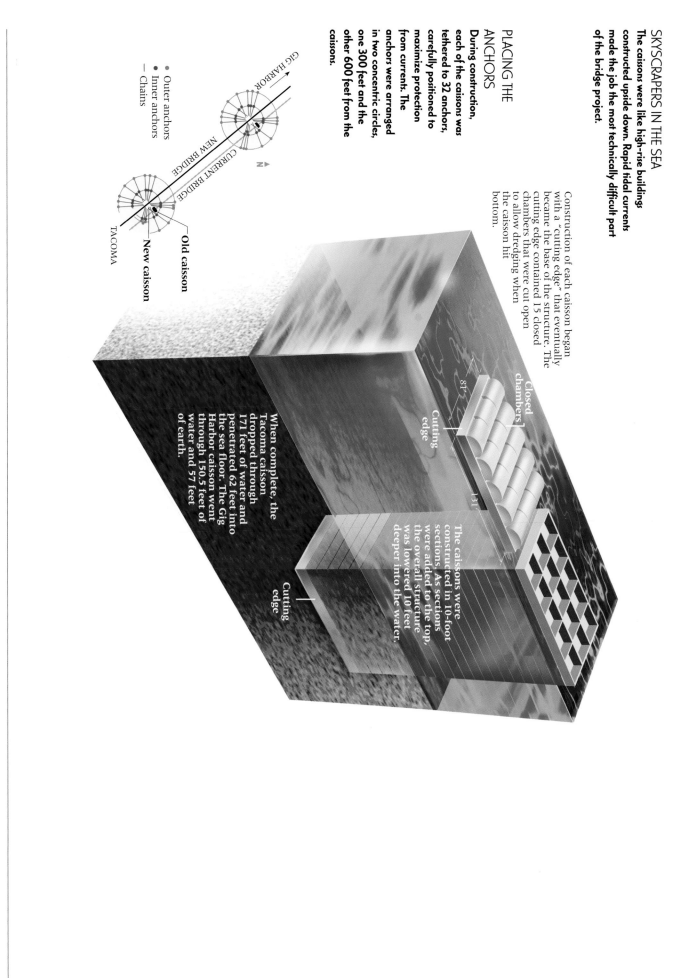

Closed chambers

Cutting edge

81'

131'

The caissons were constructed in 10-foot sections. As sections were added to the top, the overall structure was lowered 10 feet deeper into the water.

Cutting edge

When complete, the Tacoma caisson dropped through 171 feet of water and penetrated 62 feet into the sea floor. The Gig Harbor caisson went through 150.5 feet of water and 57 feet of earth.

water ridge. That means the passage narrows not only horizontally but vertically as well.

"It's an incredible Mixmaster," said Curtis Ebbesmeyer, a University of Washington tidal expert whose consulting company helped TNC analyze the currents. "When the bottom water hits the ridge it rushes to the surface like an upward waterfall. It climbs 600 feet, about the height of the Space Needle, every flood tide."

The complexities of Narrows' currents are clearly visible from the air. When the tide is running, the water surface is alive with ornate scrolled patterns, writhing and uncoiling as they rush downstream.

That the new bridge was to be built so close to the old one complicated Sherman's problems. On incoming tides, the water hit the forward edges of the existing caissons and created powerful vortexes as it passed. The water on their leeward sides—exactly where Sherman would be trying to keep the new caissons in place—boiled like big Jacuzzis.

Sherman's task was further complicated by the necessity of protecting the wreckage of Galloping Gertie, which lay scattered on the sea floor beneath the construction site. The old bridge was listed on the National Register of Historic Places and was not to be disturbed.

Over the years, the failed bridge also had become an unofficial but sensitive environmental preserve. Before Gertie fell, the floor of the Narrows was swept clean by currents and supported virtually no marine life. The tides whisked all the fine material away, leaving a surface that resembled a cobblestone street. Boulders propelled by the tides clattered through the passage like bocce balls.

When the bridge fell in 1940, it created refuges for marine creatures. They quickly came to colonize, thriving on the fresh supply of algae, plankton and

krill that arrived with every tide. Anemones, barnacles, mollusks and mussels attached themselves to twisted girders. Record-sized octopuses, lingcod and rockfish took up residence behind broken concrete. Crabs, wolf eels and sea stars settled in, rounding out an unusually lush ecosystem.

Scuba diving guidebooks list Gertie's wreckage as a prime destination for advanced divers, who are able to descend beyond 60 feet in the brief slack periods between tides. Puget Sound divers, led by a particularly avid Gertie guardian named Robert Mester, were determined to protect Gertie's remains from construction damage.

Gertie's listing on the National Register was primarily due to a personal crusade by Mester. When it became clear that building the new bridge would have at least some minor impacts on the wreckage, he regarded the threat as something close to sacrilege.

"It's outrageous," Mester said. "If somebody tried to do the same thing with the USS Arizona in Pearl Harbor, they wouldn't be allowed to do it. They'd have to figure out some way around it."

As the project moved forward, the divers made it clear they would be watching Sherman's every move.

PLANNING AND THE WITCHES

During 2001, long before any actual construction took place in the Narrows, Sherman and the TNC design team spent thousands of hours in meeting rooms, running calculations and planning.

Manuel Rondón, an unrelenting and meticulous planner himself, wanted every contingency examined. His prescription for preparing for complicated projects went like this: First, think of everything you can that might go wrong and figure out what to do if it does. Then come up with everything you don't

▶ A giant lingcod peers out from shelter created by the wreckage of Galloping Gertie. Colonization by marine animals turned the remains into a lush marine reserve.

▲ Robert Mester, instrumental in winning federal protection for Gertie's remains, fought to protect the old bridge from new construction.

▲ Near the Gig Harbor shore, diver Russ Steffensen checks the wreckage for construction damage.

think will go wrong, but by some miracle might, and figure out what to do in each of those cases.

After that, all that remained were the things no one could imagine, the ones that came out of nowhere and took you by surprise. Those variables Rondón called "the witches behind the door." The mark of true competence in construction, he said, was being able to react calmly and creatively when one of the witches appeared.

Planning for caisson construction was made easier by engineering field notes left by Galloping Gertie's builders. The essential process of building the caissons had not changed significantly since then, though the new caissons needed to be bigger and stronger.

In 1939, engineers had solved the problem of anchoring the caissons with a blunt-edged but effective solution: They dropped concrete blocks as big as mobile homes in circles around the caissons and linked the caissons to them with chains.

The big blocks, 49 of them, each weighing about 600 tons, were poured at Tacoma's City Waterway and hauled to the Narrows on barges. Dropping the anchor blocks was a big crowd pleaser in 1939. Hundreds of people gathered on the bluffs in Tacoma and Gig Harbor to watch the barges rear up like teeter-totters and the great splashes when the blocks hit the water.

At first, TNC considered using concrete blocks as anchors, too. But Sherman remembered reading about a new anchoring system that had been used by the U.S. Navy and on a few offshore oil rigs. If it would work in the Narrows, it would be considerably cheaper and more accurate than concrete blocks. One day he brought a drawing in to work.

The anchor he proposed was basically a steel spear driven into the sea floor with a chain attached at its balance point. When the chain was pulled, the shaft turned parallel to the surface, locking it into place.

"To come out they would have to lift out the entire cone of soil above them," Sherman said, demonstrating by making a V-shape with his hands. "The deeper they go, the harder they are to pull out."

Welders built prototypes using 8-foot lengths of steel I-beams, and TNC tested them in the Narrows, driving them in with vibrating hammers combined with high-pressure water jets to loosen the sand and gravel. Driven about 50 feet deep, the anchors resisted 500 tons of force, 100 tons more than they were expected to encounter during construction.

Where to place the anchors was another matter. They needed to be configured so they shared the load of the caissons equally at all tides throughout the construction process. The solution in 1939 had been to arrange the anchors symmetrically around the caissons, with the chains arrayed like spokes from the hub of a wheel. That wouldn't work this time because the caissons on the existing bridge disrupted the currents so much.

"The old caissons change the flow regime completely," said Brenda Lichtenwalter, the TNC engineer who coordinated the anchoring system. "We're sitting right there in these vortices. It's a really confused flow pattern."

To figure out where to place the anchors, Lichtenwalter used data from two modeling tests. The first was a virtual model, which used computer software to predict the direction and intensity of forces. Aided by ultra-sensitive sonar readings that tracked the movement of plankton and other tiny bits suspended in the water, software engineers subdivided the moving water into millions of theoretical cubes, then combined them to create pictures of how the water moved.

The other model was a real one, a 1/100th-scale replica constructed by the British hydraulics research firm, H.R. Wallingford Ltd. A team of modelers at the firm's testing laboratory in Wallingford, England, recreated the bathymetry of the Narrows in a flow tank 100 feet long and 30 feet wide. They then inserted 2-foot-tall replicas of the caissons, old and new, and wired them with tiny strain gauges.

Lichtenwalter observed the tests, watching from a control room as technicians waded through the tank in hip boots, making minute adjustments. Lasers aimed at the caissons measured movement in

▶ Enormous blocks of concrete used to anchor Galloping Gertie's caissons (shown as black rectangles on this underwater map) complicated the task of positioning anchors for the new caissons.

▲ TNC drove anchors for the caissons deep into the sea floor with vibrating hammers and high-pressure water jets.

each of the six degrees of freedom of submerged structures: sway, surge, pitch, yaw, roll and heave.

Based on data from the two models, TNC came up with an arrangement of 32 anchors for each caisson, positioned at varying distances along the arcs of two concentric circles, one 300 feet from the caissons and the other 600 feet. The lines from the inner circle would be attached near the bottoms of the caissons. The lines from the outer circle would attach near the tops with a moveable attachment that could be adjusted upward as new sections were added.

In the Narrows, crews guided by Global Positioning System locators hammered 16 anchors into the perimeter of each circle, sometimes altering the plan slightly when they encountered Gertie's remains. They linked the anchors to laughably large chains, each link as big as a coffee table and weighing 240 pounds. They then tied the chains—some more than 600 feet long—to tethers that ran to floats on the surface, ready for the caissons to arrive.

CUTTING EDGES AND THE BIG FLOAT-OUT

As crews at the Narrows installed anchors, welders at Todd Shipyards in Seattle worked on land to create the caisson bottoms, called "cutting edges" because they would slice into the sea floor. Because of the tremendous force they would have to withstand, the cutting edges needed to be nearly indestructible. They were made of thick sheets of steel and welded into 15 airtight cubes, each 20 feet square and topped with a red steel-domed roof.

When the cutting edges were finished, they were as big as department stores and, as Transportation Secretary MacDonald put it, they looked like enormous versions of the boxes that Christmas balls come in.

Todd prepared to launch the cutting edges the same way it launched ships, sliding them down long, greased runways into Puget Sound. The bridge builders regarded the first launch as a symbolic moment in the history of the project: After years of planning, the first bit of the new bridge would finally hit the water.

On launch day, March 17, 2003, about 50 people gathered in the predawn darkness for the occasion. They stood watching, all wearing hard hats and safety glasses as Linea Laird, the state's project manager, and Judy Poloai, a welder at Todd for 25 years, together swung a bottle of champagne into the steel wall of the structure.

"On behalf of the people of the state of Washington, we christen thee Caisson K1," Laird said as she swung the bottle. The name, though not evocative, was technically accurate. It was the way the structure was identified on engineering drawings.

When the caisson hit the water, Rondón hugged Laird so enthusiastically her hard hat fell off.

Rather than heading straight to the Narrows, the caissons went first to the Port of Tacoma's Sitcum Waterway, 31 miles away, where TNC crews reinforced the cutting edges with concrete and built 60-foot-tall concrete walls on their tops.

On July 21, 2003, when the first caisson left the port for the Narrows, people crowded onto the decks of restaurants on Ruston Way and turnouts in Point Defiance Park to watch. The caisson moved in a sedate procession that included four heavy Foss tractor tugs, two of which had been diverted from oil tanker escort service in the North Sound, a Pierce County sheriff's marine escort and two TNC assist boats. They moved at a barely perceptible 1.5 knots out of Commencement Bay, around Point Defiance and into the Narrows.

Chuck Proulx, 81, of University Place, staked out a picnic table at Owen Beach in Point Defiance Park to watch the caisson move past. Proulx said he felt a special affinity for the new bridge because when he was a young man, he had seen Galloping Gertie fall.

"The wind was blowing and we heard some rumors, so me and two friends skipped out of school," Proulx said. "We got there in time to see it hit the water."

"I'm going to follow this all the way through today," Proulx said as the new caisson floated by. "This kind of adds to your life, to see a little bit of everything."

By the time the first caisson arrived in the Narrows, a swarm of spectators' boats, from kayaks to cabin cruisers, surrounded it. People lined the sidewalk on the old bridge, peering over the edges and slowing traffic.

As spectators watched, the tugboats deftly pushed and pulled the caisson into position between two work barges anchored next to the Tacoma tower. The first section of the new Narrows bridge was officially in place.

THE FIRST WITCH: TANGLED ANCHOR LINES

The first witch appeared shortly after the crowds had gone home. Sherman's plan was for crews to haul the anchor lines over from the floats they had been attached to weeks before and hook them to the caisson while the tugboats held it in place.

He figured the tugs would be necessary only until eight anchor lines had been fastened, the minimum considered necessary to hold it.

The caisson had arrived as planned that Monday at 10 a.m. Sherman estimated he could have the

▲ **Workers at Todd Pacific Shipyards in Seattle put the finishing touches on one of 15 domed chambers at the bottom of what would become the Gig Harbor caisson. The heavily reinforced caisson bottoms were called "cutting edges," because they would be the portion of the giant structures that would slice into the sea floor.**

Air pressure in the watertight chambers helped the caissons float and could be individually adjusted to help steer the caissons straight toward their targets on the bottom.

eight anchors hooked up in 24 hours and send the tugs home.

Crews attached two anchor chains with no problems. But then, as the tugs sat revving their big diesel engines against the tide, the crane operator made a puzzling discovery. The next anchor line could not reach the caisson. Sherman sent divers down to see what was the matter, and they came back with bad news. The back and forth movement of the tides had tangled several of the heavy chains, tying them in knots and entangling them in their tethers.

The only option, Sherman decided, was to keep the tugs in place and begin the time-consuming process of untangling the chains.

Using two floating cranes and a team of 20 divers, Sherman's crews worked around the clock, laboriously twisting and turning the chains to work out the kinks. Nobody at TNC pressured him during the process, Sherman said. "I was given tremendous latitude, even though everybody knew we were out there for millions." Still, one senior engineer did a rough calculation and took every opportunity to needle Sherman with it: keeping the tugboats and extra crews in place was costing about $100 a minute.

After seven days, workers had the necessary eight anchors hooked up and the tugs could safely leave.

In his 40 years on the water, Sherman had learned that, aside from tides and currents, the single most consistent problem in marine construction was providing access to the work site. The Narrows job was no different. The caissons were anchored a quarter-mile from the shore, and crews would have to spend more than a year building concrete walls and rebar on their tops.

"We had 100-plus people on day shift and 50 on night," Sherman said. "My problem was, 'How in the

heck do I get 100 guys out there and have them be productive?'"

This time, the proximity of the old bridge turned out to be an advantage. Carpenters built walkways beneath the deck of the old bridge, running from each shore as far as the towers. Then they attached flights of metal stairs that dropped 210 steps to the base of the towers. From there, workers could cross the 60 feet over to the new caissons on gangplanks designed to rise and fall with the tides.

The old bridge also came in handy for supplying concrete. Sherman set up big electric pumps at both ends and ran supply pipes along the walkways. Concrete trucks loaded up at a temporary batch plant on the Gig Harbor side and delivered their loads to the pumps like participants in an old-fashioned fire brigade.

The flat platforms at the bases of the old towers also made handy locations for a field headquarters.

▲ Avid bridge watchers at Tacoma's Point Defiance Park keep an eye on the first section of the Tacoma caisson as tugs haul it slowly out of Commencement Bay and into the Narrows.

▲ With 47 feet of their 78-foot depth under water, the caisson bottoms made unwieldy loads for Foss tugboats. Before taking the first caisson into the Narrows, where they would encounter heavy currents, tugboat operators practiced maneuvering the massive structures in Commencement Bay.

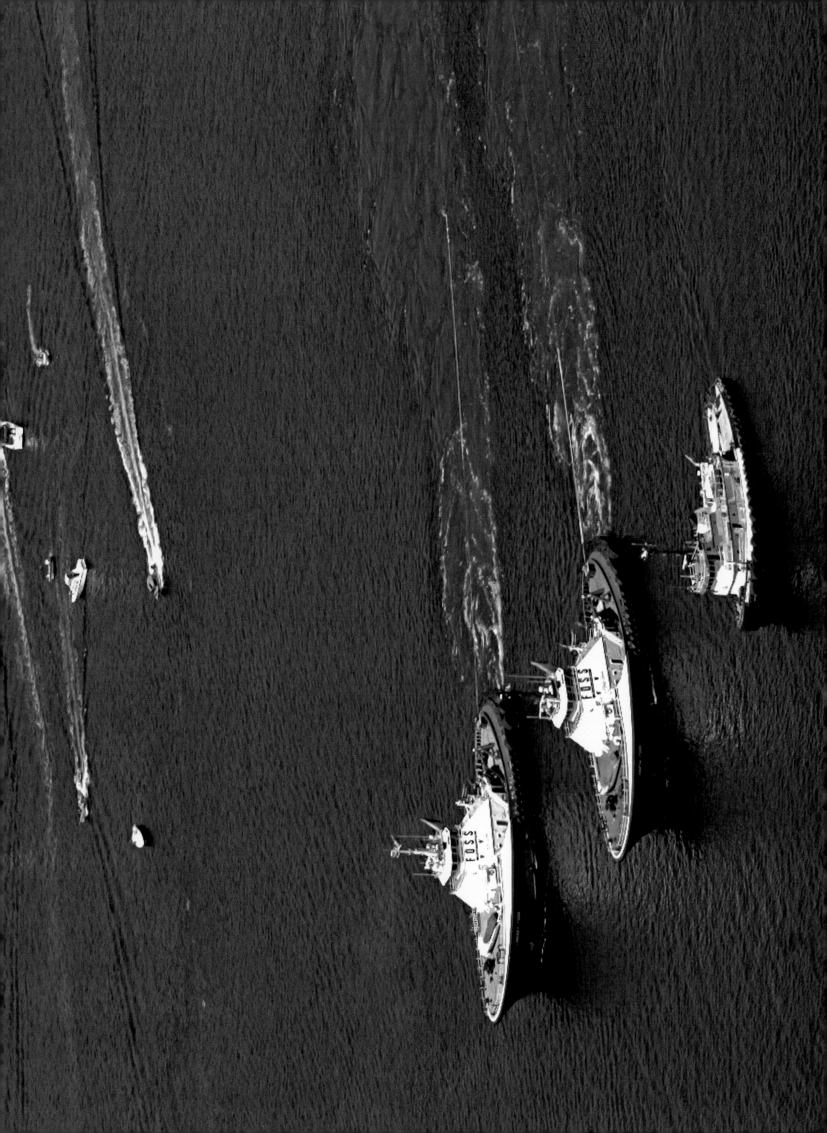

TNC barged in 18 steel shipping containers and arranged them in two neat stacks connected by stairways and balconies.

The windowless containers, each 8 feet wide, 8 feet tall and 24 feet long, were equipped as lunchrooms and changing rooms for the crews, tool storage containers, offices, conference rooms and even a bunkhouse, set up with a couple of cots and sleeping bags for emergency sleep-overs.

The two container communities became known as Piertop City East, on the Tacoma side, and Piertop City West, on the Gig Harbor side.

Lichtenwalter set up an office there to monitor the anchorage system, in a container crammed with desks and computers, plus a microwave oven, a coffee pot and a refrigerator.

"It's not exactly home," she said, "but I've worked in places that were set up a lot worse."

The engineers shared the piertop cities with an exceptional crew of construction workers. Ironworkers, carpenters and laborers at local unions all were eager to work on the bridge project, giving Sherman the luxury of being able to pick and choose. "They are the cream of the crop," he said. All were expert builders, but some had never done any marine construction, or as Sherman put it, "wet work."

"It was a new thing for a lot of people," he said, "leaving work one day and finding the job in a different position the next day, having to keep things tied up for the tide changes, just knowing that you're not on terra firma."

For some it was a big transition, but it didn't bother Todd Reising, the foreman of a crew of ironworkers who assembled the massive webs of reinforcing steel that needed to go into the caisson walls. At 33, Reising was on the old side for an iron-

▲ PREVIOUS PAGE: Surrounded by a swarm of sightseers in private boats, tugboats lead a historic procession: the arrival of the first section of the new bridge in the Narrows. Traffic on the 1950 bridge slowed to a crawl as the base of the Tacoma caisson passed underneath.

▶ To provide access for workers and engineers, TNC designed and assembled temporary field offices made of steel shipping containers on the piertops of the 1950 bridge. Called Piertop City East and Piertop City West, they included offices, lunchrooms, conference rooms and sleeping areas for emergency overnights.

worker, but he was built like a chunk of iron himself and wore his 45-pound tool belt as easily as a sweater vest. In his off time, he played the guitar and sang for a heavy metal band called Idlemine, where he thrashed out chords with the same intensity that he put into hammering rebar.

Building a caisson was in most respects not that different from building a high-rise building, Reising said. The sequence was the same in both: Set up the removeable forms used to hold the wet concrete in place. Assemble a web of steel reinforcing rods inside the forms. Pour the concrete and let it set.

► **Getting to work meant traversing a quarter-mile walkway beneath the old bridge, on the left, descending 210 steps and then crossing a gangplank, lower right.**

manhandling big concrete-supply hoses on booms, spreading concrete and then using hydraulic vibrators to make sure there were no voids.

The pace was relentless, both during the pours and in setting up the reinforcing steel. "In steel, the idea is to put in as many tons in a day as you can possibly put in," Reising said. "To do that, you have to get in an aggressive, no-holds-barred state of mind."

Weather conditions and currents in the Narrows also made the bridge job different. The big grids of crisscrossed rebar were put together on shore and then hoisted into the caissons by crane. When the wind blew, the rebar grids acted like sails. Wake from passing boats bounced the steel up and down as ironworkers tried to wrestle it into place, putting them at risk for severed limbs.

▶ Ironworker foreman Todd Reising, helping his crew construct the heavy web of steel reinforcement that would be encased in the concrete walls of the Tacoma caisson.

"In steel," Reising said, "The idea is to put in as many tons in a day as you can possibly put in. To do that, you have to get in an aggressive, no-holds-barred state of mind."

When finished, each of the caissons would contain approximately 2.9 million pounds of reinforcing steel.

BENEATH THE SURFACE

Construction workers on the caissons were often joined by teams of divers, as many as 20 on the job at once.

Most of the divers came from Washington and Oregon, a few came from as far away as California and Arizona. Many had dived together all over the world, from Trinidad to the Arctic, working in gas fields, on oil rigs and dams.

At the Narrows, the divers worked both inside and outside the caissons, helping adjust the anchor chains, installing instruments and facing many risks. Outside meant dealing with the currents; inside, gases from cutting torches could accumulate in hidden air pockets and suddenly explode. Pressure differences between the air chambers in the cutting edges and the shafts above had the potential of sucking divers into holes or blowing them to the surface at speeds that would burst their eyeballs.

Remove the forms. Start over.

"That's basically a high rise right there," Reising said, pointing at the Tacoma caisson. "It's a skyscraper in the sea."

One difference was that each of the 10-foot "lifts" added to the caissons had to be accomplished in a single continuous session of pouring concrete— an average of 1,270 cubic yards each time. Letting the concrete set up and then pouring wet concrete on top of it would make a weak joint. What that meant was that, once ironworkers had the forms and rebar set and the concrete started flowing, it could not end until the lift was finished. Equipment malfunctions or other delays sometimes kept crews on the job for 12 or 14 hours at a time. They spent the time

To avoid the bends, the divers were able to work only about an hour at a time on the bottom. They then had to surface slowly, riding in a cage called a "man basket," hoisted by a winch, pausing every 10 feet to allow their bodies to readjust to the reduced pressure.

When a diver surfaced, he had just 38 minutes to scramble out of his suit and into the decompression chambers set up on the tops of the caissons.

"In diving, you don't make too many mistakes," said Kerry Donohue a big, good-natured Canadian diver with prematurely gray hair.

A half-century earlier, the divers who worked on Gertie's caissons did not fare well. All survived the work in the Narrows, according to ironworker and historian Joe Gotchy, but one was paralyzed less than a year after the bridge was finished. Another had a cerebral hemorrhage, and a third died when his diving suit ruptured while diving at Pearl Harbor after the Japanese attack.

Better equipment and a better understanding of physiology had removed many uncertainties in the diving profession since then, Donohue said. Despite the dangers that remained, he said, he and the other divers looked at the Narrows project as a plum job.

"Everybody knows about Galloping Gertie," he said. "The guys like to work on it just to say they worked on the bridge."

Part of the attraction for all workers was the natural setting. On the bottom, divers sometimes played gentle games of tug-of-war with octopuses attracted to their tools. Up on top, the piertop cities were idyllic spots in good weather. When they had time to notice, workers could see salmon, seals, giant jellyfish, porpoises and even orca whales.

One day crews on the Gig Harbor caisson spotted something in the water they could not identify. It

seemed to be struggling, paddling first one way and then another in the current in the middle of the channel. A foreman radioed Kent Lowe, captain of 68-foot tugboat Redwood City, which TNC had chartered to serve as the piertop cities' taxicab, delivery van and emergency assist vehicle.

Lowe set out in the direction of the splashing. When he steered the Redwood City in close he saw that it was a young deer, disoriented and exhausted. Lowe chugged over to the tired animal and eased the tug alongside. His deck hand lassoed it and hauled it aboard.

"It only weighed about 70 pounds," Lowe said. "We took it over to the beach on the Gig Harbor side and let it go."

The caissons were pleasant in the summer, but as construction continued into winter, when icy winds blasted in from the southwest and rain blew sideways, they sometimes seemed like remote Arctic outposts. During the two daily work shifts, the piertop cities were highly charged centers of energy and noise. At night, Lichtenwalter and a couple of other anchor specialists were usually there by themselves, monitoring the anchors. To make sure none of the anchors was taking too much of a load, strain meters had been incorporated into every anchor cable. Once each second, data from the strain meters was relayed to computers in the piertop cities. Lichtenwalter and other engineers monitored the anchors 24 hours a day, seven days a week, watching the changing numerical values on computer screens.

It was lonely work, and usually quiet except for the foghorns that blared back and forth across the Narrows in an earsplitting call and response.

The job grew more difficult as the surface area of the caissons increased.

The steel cable connecting the chain to adjusting pulleys had broken, meaning the entire adjustment system needed to be removed, rebuilt and replaced.

"It was a rigid connection and when it moved back and forth with the tides, it stressed the steel," Sherman said.

Between Thanksgiving and Christmas, a total of five connectors broke loose, four on the Tacoma side and one on the Gig Harbor side. Knowing that if those connectors were stressed to the point of breaking, others undoubtedly were too, he quickly doubled up the connectors on all the chains with arresting straps.

TOUCHDOWN!

From the very beginning of caisson construction, the critical moment of concern was the moment of touchdown, when the giant towers would be lowered into place on the bottom.

To ensure proper alignment of the towers, the caissons had to be planted precisely in their prescribed footprint, with a margin of error of just inches in any direction. Because the floating caissons would at that point be as tall as 15-story buildings and weigh 33,000 tons, dropping them accurately would not be a simple matter.

For the crews and engineers, it was a moment like the moon landing. Months of planning preceded the complicated maneuver. Work crews and crane operators spent a week rehearsing and working through possible problems. The Gig Harbor-side caisson was ready first because the water was shallower on that side of the channel. December 15 was chosen as touchdown day because of a nice long, low tide at 4:15 p.m. that would lower the caisson to just 8 feet from the bottom.

▲ Concrete pumped from the shore was distributed into caisson walls with long booms, articulated like insect legs. Beneath the surface, anchors strained to keep the growing caissons in position.

▶ On top of the caissons, workers extended steel and concrete walls in 10-foot-tall increments, then lowered the structures deeper into the water.

One night in Piertop City East, as the engineers watched the monitors during a heavy current, they saw the load for one of the anchors suddenly drop to zero. They thought it must be a computer glitch or a bad strain meter. In fact, it was the second witch.

If the caisson were going to break loose, this was how it might begin. Keeping the caissons in place depended on all of the anchors working together. If enough of them failed, it would overstress others and could start a chain reaction.

"As soon as one went to zero we knew something was happening with it but we didn't know what it was," Sherman said. A hasty underwater inspection revealed that the anchor chain had broken free from its attachment at the caisson.

TNC publicly downplayed the event, but top managers at the Bechtel and Kiewit corporation headquarters paid close attention, dropping in frequently for visits in the days before.

The plan was to lower the caisson by a combination of means. Air would be released from the 15 compartments in the cutting edge, decreasing buoyancy. Each compartment had an individual valve, meaning technicians could raise and lower or tilt the caisson by choosing the right combination of compartments. Meanwhile, water pumps capable of pumping 5,000 gallons of water a minute had been set up on top of the caisson to add sea water and increase its weight.

To monitor the caisson's position as it descended that day, engineers established several points of reference, wanting to make sure it was correctly aimed and not twisted or out of plumb. GPS units triangulated data from three satellites. Laser beams shot from shore to the center of the caisson made sure it was properly aligned. To triple check the position, an electronic distance meter on the foundation of the existing bridge shot more laser beams at reflectors on the four corners of the caisson, delivering distances in fractions of millimeters.

The planning for the touchdown was so detailed that engineers even calculated the probable decibel level of the whoosh of rushing air when it was released from the bottom chambers. They gave serious consideration whether to install mufflers on the air valves or have workers use ear protectors. (In the end, they went with the ear protectors.)

On the big day, Sherman stationed himself on a work barge anchored next to the caisson rather than with the cluster of engineers gathered around computers in Piertop City West.

In addition to all the electronic monitoring equipment, the caisson had simple depth markings painted on its sides. Sherman preferred to watch those. "It took too long for all the surveyors' data to be crunched through the computers," he said. "I don't like to wait for all that."

In the end, the touchdown went so well it seemed anticlimactic. The water remained calm, all the gadgets worked as planned and the caisson gently snuggled into position just 8 inches south and 9 inches west of dead center—well within the limits.

As soon as the caisson touched bottom, workers switched on the pumps, dumping 50 vertical feet of water into the wells and planting the caisson firmly in place against the rising tide.

Sherman remembered a good feeling, but not an impulse to jump up and down with joy.

"The completing of it, the security of it finally setting on the bottom, of knowing that it's in location is a relief," he said. "But after all the planning and calculations, when it actually happens out in the field, it's what you expect to happen."

The touchdown on the Tacoma side went even better, even though the water was deeper on that side and the currents stronger. On January 15, Caisson K2 landed ¼ of an inch south and 9 inches east of the bulls eye.

The successful touchdowns were the climax of the caisson construction drama, but they did not mark its end. The steel tops on the cutting edge compartments still had to be removed to provide access to the sea floor. When the tops were off, cranes equipped with clamshell buckets dug sand and gravel out of the bottoms of the wells, sinking the caissons deeper as crews continued to add to their tops.

Both caissons were finished in August 2003 and ready for tower construction. From the surface, the only evidence of 23 months of labor was two smooth cubes of concrete, riding slightly above the water next to the old bridge.

"In marine construction most people don't see the things you build," Sherman said. "They won't see this either. But this is the whole foundation for the whole thing.

"I don't call it a monument to anything," he said, "but it's something that will be there for the foreseeable future, and that's a good feeling."

▶ The final touchdowns were guided from mission control offices in the piertop cities. The shipping container offices received data from three GPS locators as well as electronic distance meters.

▲ As the caissons neared the bottom, workers continued to add sections on top.

"To closely duplicate the design of the existing bridge would be to diminish its uniqueness and historical significance," the director of the Washington State Historic Preservation Office advised planners. "The new bridge should have its own distinctive character and reflect its own time."

Freedom from the past seemed to give designers artistic license to make bold architectural statements with the new bridge's towers and anchorages, its most prominent features. Tower design in particular establishes overall character—the cathedral-like Gothic arches of the Brooklyn Bridge, for example, and San Francisco's art deco Golden Gate, with its stepped tower tops and red-orange glow.

However, designers of the new Narrows bridge faced many other constraints. Because the old and new bridges would stand so close together, they needed to work as a visual pair, not competing but complementing. In the relationship, it was generally agreed, the new bridge was to be the bridesmaid, not the bride. Other pressures also pushed for a conservative, utilitarian approach. The tower height, a critical starting point in suspension bridge design, was restricted to the 510-foot height of the old bridge, partly out of respect for the past but also because taller towers raised safety concerns for pilots at the Tacoma Narrows Airport, less than a mile away.

As a primary building material, reinforced concrete was so practical and relatively inexpensive it made other options seem frivolous in comparison. Steel towers would have cost twice as much. Concrete may be bland, but it is easy to take care of, a big plus for the state Transportation Department. Concrete requires almost no maintenance, while steel bridges in saltwater environments keep maintenance crews busy year-round, scraping and repainting. In 2005, maintaining the old Narrows bridge took a crew of six and cost $850,000.

For their part, state traffic engineers were mostly interested in moving vehicles safely and efficiently. They didn't want the new bridge to be a visual distraction for drivers and even went so far as to suggest raising the side rails so people wouldn't have a water view as they crossed.

The fact that the new bridge would be paid for with tolls was perhaps the strongest inhibiting influence of all. No one was in the mood to pay for frills. Politics and public funding controlled the design environment, and in that setting, function easily trumped form.

In the end, the look of the new towers and anchorages was determined almost totally by cost and function. The anchorages, which would hold down the main cables at either end of the bridge, would be innocuous boxes designed to disappear beneath the highway. The tower design included only two purely aesthetic touches. The cross braces that connected the tower legs would be marked with double X's imprinted in the concrete, a nod to the prominent crossed steel struts in the existing bridge. And the tower pedestals, rectangular blocks of concrete at the base of each tower leg, would be beveled at their tops rather than square because they looked more streamlined that way.

A third change was considered but rejected, according to Tim Moore, the Transportation Department's authority on bridges and the man in charge of reviewing the builder's designs. In drawings, it appeared that moving the middle struts closer to the tower tops would make them more graceful. The state design team considered that, Moore said, but in the end decided against it

because it would have reduced the towers' strength and made them more difficult to build.

ANCHORAGES

Aesthetics aside, a suspension bridge's anchorages have a very simple job to perform: they hold down the ends of the main cables. Bulk is the main requirement, but building them is more complex than simply digging holes and filling them with concrete.

At the Narrows, location and cost restrictions meant the anchorages would have to rely on shape as well as bulk. Designers needed to find a shape that would maximize the anchorages' holding power by

▶ Deep inside the earth, an ironworker readies rebar for the next concrete pour on the Tacoma anchorage. When finished, each anchorage weighed 81 million pounds.

▲ While sheer bulk was necessary to resist the force of the main cables, the anchorages also rely on shape to do their jobs efficiently. Here, ironworkers prepare a wall inside a 63-foot-deep excavation near the rear of the Tacoma anchorage designed to help keep it from sliding forward.

putting their bulk precisely where it was needed to resist forces put on them by the cables. In this case, they would have to resist 25,000 tons, a force equivalent to the weight of a small aircraft carrier.

The process began with some nonnegotiable givens. Because the new anchorages would stand right next to the old ones, their width was limited. Also, the anchorages needed to be designed as drive-through structures to keep open the option of someday adding a second, lower deck to the bridge.

The depth of the water table on shore also directed the design. Water can act as glue in soil, but it also can act as a lubricant. The anchorages had to be built above the water table to avoid landslides.

"The question was not only, 'Is the bridge going to move the anchor?'" Moore said. "It was, 'Is it going to try to pull the entire hillside apart?'"

The groundwater on the Gig Harbor side was more than 100 feet from the surface, deep enough so it was not an issue. But on the Tacoma side it was just 70 feet down, which restricted the structure's depth.

To make sure the force of the bridge wouldn't pull the anchorages out of the ground, designers had to determine how much friction the structures would create against the soils surrounding them and come up with a shape that would put the least stress on weak areas. They began by boring 3-inch-wide holes 150 feet deep at the anchorage sites and pulling up core samples. The samples showed bands of silt, sands and clays, some firmly compacted by the 3,000 feet of ice that covered the South Sound during the most recent glacial advance 15,000 years before. Other samples were loosely compacted and had practically no strength.

Computer simulations mathematically subdivided the earth into thousands of cubes, each with its own physical properties and susceptibility to stress. The individual predictions were then combined to create three-dimensional maps that indicated how the structures would react as a whole and revealed potential "slip planes," where landslides were most likely to occur.

In addition to calculating the amount of force the cables would exert, the designers needed to know the precise direction of the force. Because the cables would be attached to the rear of the anchorages and pull forward and upward, they would tend to push the anchorage's forward edges down and lift the rear ends. For that reason, the backs of anchorages were designed to be deeper than the fronts. In profile, the anchorage plans looked like maps of Texas, with large humps in the rear. The irregular shape had another advantage: it fit into the earth like a key in a lock, making it less likely to slide forward.

The anchorage design was complex, but once the plans were finished, construction would be mostly straightforward physical labor, a process not so different from building the caissons: building forms, installing rebar and pouring concrete.

In October 2003, just before the first concrete was to arrive at the excavation for the Tacoma anchorage, Flint Gard climbed into the chasm, checking last-minute details. At 116 feet across and 63 feet deep, the hole was so big it appeared to be some sort of natural geologic feature. He needed three sets of ladders to descend to the deepest part of the excavation, at the point farthest from the water.

In three months, Gard would be in charge of building the anchorage's twin on the Gig Harbor side, but for the time being, he was helping coordinate crews on this side of the water.

▲ Flint Gard was the TNC engineer in charge of building the Gig Harbor anchorage. "Just the sheer size of it makes it challenging," Gard said of the job. "But it's really pretty much a logical progression of steps."

"Just the sheer size of it makes it challenging," Gard said, "but it's really pretty much a logical progression of steps. It's like what they say about eating an elephant. You just have to take it one bite at a time."

Like Tom Sherman, Gard was a Washington native with a long family history in the Puget Sound region. His grandparents settled near the little waterfront settlement of Olalla in 1908. At 42, Gard had managed several big construction projects along the West Coast and in Hawaii, but never anything as big as the anchorages.

"This kind of thing only comes around once in a while," he said. "There are not too many people who can say they've done it."

In January 2004, the lower sections of the Tacoma anchorage were filled with concrete, and the Gig Harbor excavation was ready for its first pour. Even though the Tacoma workers had a three-month head start, Gard and his crew challenged them to a race to see who could finish first. The Tacoma side, headed by Steve Chambers, had a big advantage, but Gard expected to gain time by watching how they did things and figuring out ways to do them faster.

Glacier Northwest's blue and white concrete trucks made thousands of trips to the two anchorages through 2004 and into the spring of 2005. Gard's crew gained steadily but never managed to catch up. The Tacoma side finished in April, the Gig Harbor side in May. While Gard's crew came in second, they took pride in the fact that they had taken less total time—just 14 months compared with 18 for Tacoma.

When the anchorages were finished and earth filled in around them, there was little indication of the size of what lay beneath the surface. With 95 percent of the structure underground, all that

was visible were unadorned gray boxes that, with proper signage, could have passed for banks in a suburban mall.

Thanks to their design, though, they were models of efficiency. On a strength-per-pound basis they far outdid the anchorages on the 1950 bridge. The old anchorages weigh 132 million pounds each. The anchorages for the new bridge came in at a svelte 81 million pounds, even though they would support a bigger bridge.

According to engineers, the new anchorages had a built-in safety factor that made them capable of holding down nearly twice as much force as the 25 million pounds each cable would exert.

And they would barely budge doing it. When the full weight of the cables and deck were applied, engineers predicted, the anchorage at the Gig Harbor end of the bridge would move just three-fifths of an inch horizontally and three-quarters of an inch downward. The Tacoma anchorage would move even less—a mere two-fifths of an inch in each direction—about the width of a ballpoint pen.

TOWER DESIGN

Suspension bridge towers have heavy burdens to bear. The new Narrows bridge towers would have to carry the 12 million-pound cable system, plus the 53 million-pound deck suspended from them, as well as all the cars and trucks on the bridge at any one time. Most of that weight would be transferred into downward force at the tower tops.

The towers needed to be strong enough to resist crumbling under that tremendous downward load but supple enough to twist and sway with the infinite combinations of forces that would push at them from their sides—movements ranging from subtle

▼ A laborer steadies a supply pipe feeding wet concrete into the Tacoma anchorage. Unlike the first two Narrows bridges, women worked alongside men in nearly every aspect of construction.

▲ Concrete provided most of the
anchorages' bulk, but they also
required massive amounts of
reinforcing steel—about 1 million
pounds for each anchorage. According
to designers, the steel would help them
withstand earthquakes rated as high
as 9.1 on the Richter scale, more than
10 times the magnitude of the earthquake that leveled San Francisco
in 1906.

The anchorages' design was complicated by the state's requirement that,
if it should become necessary, they
would be able to accommodate a
second, lower level of traffic on the
bridge. If a second level is ever built,
vehicle traffic—or possibly passenger
trains—would pass through the
opening to the right in this photograph.

contractions and expansions caused by changing temperature to the violent conniptions of earthquakes.

Wind is the most famous natural force at the Narrows, having taken the blame for Galloping Gertie's fall. The 13-mile stretch of open water south of the Narrows is precisely aligned with the prevailing winter wind, which effectively turns the channel into an enormous wind tunnel. Engineers setting out design parameters for the new bridge towers decided they should be able to withstand sustained winds of 109 miles per hour and gusts of 127 miles per hour. Winds that strong theoretically would occur just once every 10,000 years and were far stronger than the 42-mile-per-hour wind that doomed Gertie.

But earthquakes were a bigger concern than wind. The Narrows lies within striking distance of two major faults, the Tacoma Fault and the Seattle Fault, as well as the subduction zone off the Washington coast where the Juan de Fuca plate dives under the continental shelf. All were capable of sending powerful and complex waves of motion through the bridge—unpredictable combinations of rolling, heaving and lurching that were difficult to design for.

Motion from earthquakes would begin at the bases of the caissons, more than 200 feet below the water surface, and move upward through the towers, whipping them like the tips of fly rods. The potential effects of earthquakes made wind seem

▼ **An ironworker on the Tacoma tower holds onto his hard hat to keep it from blowing away in the wind. Wind is legendary in the Narrows, but for tower designers, it was a less significant threat than earthquakes.**

insignificant in comparison. The strongest winds were expected to move the tops of the towers 5 or 6 inches. The strongest earthquakes could swing them back and forth as much as 4 feet.

For engineering purposes, designers used a 9.1-magnitude earthquake as a theoretical maximum, the size of an earthquake thought likely to occur once every 2,500 years. (The highest magnitude earthquake ever recorded was a 9.5 quake in Chile in 1960.) The real maximum was unknowable, but engineers needed to choose a reasonable limit as a starting point. "If the bridge were designed for no risk, we couldn't build the towers big enough or put enough steel in them," said Moore, the state bridge design expert.

Concrete towers with internal skeletons of steel combined the structural advantages of both materials. Concrete readily resists downward, compressive force, while steel effectively adapts to lateral movement.

The cables stretched across the tower tops would help protect them from forces moving parallel to the bridge, so the main concern was perpendicular force. To resist tower flexing in that direction, designers chose a shape that was essentially a rectangular frame: two heavily reinforced legs connected by three horizontal struts.

Viewed from the front, the towers would widen at the bottom, like someone standing with legs slightly spread. That stance would have the advantage of greater stability, and it meant the deck could be hung straight down from the tower tops with room to spare on either side. In contrast, the legs of the 1950 bridge stand straight up and down, which puts them just inches from traffic.

The new tower legs would lean toward each other as they rose, and they would taper individually as well, decreasing from a width of 29 feet at their bases to 19 feet at their tops. That was efficient in two ways. It reduced the amount of concrete and steel needed, and it reduced the amount of stress at their bases, where they were most vulnerable.

As soon as the tower legs grew tall enough so their inward lean was perceptible, worried citizens

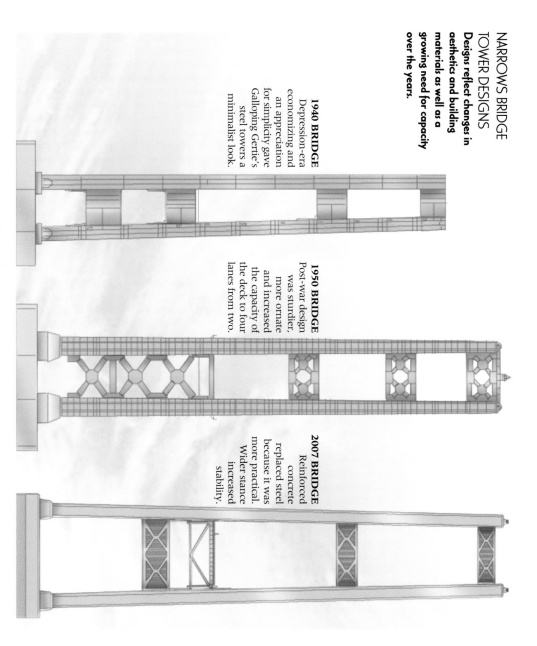

NARROWS BRIDGE TOWER DESIGNS
Designs reflect changes in aesthetics and building materials as well as a growing need for capacity over the years.

1940 BRIDGE
Depression-era economizing and an appreciation for simplicity gave Galloping Gertie's steel towers a minimalist look.

1950 BRIDGE
Post-war design was sturdier, more ornate and increased the capacity of the deck to four lanes from two.

2007 BRIDGE
Reinforced concrete replaced steel because it was more practical. Wider stance increased stability.

▲ **Bridge towers present complex engineering challenges. They must be able to withstand millions of pounds of downward force but, at the same time, accommodate the external forces of wind, temperature change and earthquakes.**

began calling the TNC office: "Do you know the towers are not going up straight?" they asked.

Erin Babbo, TNC's public relations manager, took most of the calls. "Yes, we are aware of that," she assured them. "It's part of the plan."

TOWER CONSTRUCTION

Watching the towers being built was like watching a play with the curtain closed. As the towers rose, all anyone could see were four large white metal boxes, one perched on top of each tower leg. Every few days, one or another of the boxes would crawl upward several feet, seemingly without human effort, leaving an excreted section of tower leg behind, like a discarded insect casing. Occasionally, one saw workers riding up to the boxes in the steel mesh elevators bolted to the tower legs, but signs of actual labor were rare.

The boxes, nicknamed "birdcages," were part of a revolutionary, self-jacking building system designed in Australia. The centers of the boxes were reusable concrete forms that could be moved in and out with fine, threaded adjusters. Around them, enclosed platforms provided space for workers. Tower cranes dumped huge buckets of wet concrete into the boxes' open tops and, when it hardened, the forms were cranked back. The box then jacked itself up to the next position, 17 feet higher, using the wall it just created for leverage. When the last section of a tower was finished, a crane would pluck the boxes off the top and lower them to the surface.

From outside the boxes, tower construction appeared effortless and magical. Inside, it was all noise, sweat and swinging hammers. Twelve to 15 ironworkers and carpenters crammed themselves into each box, scrambling like monkeys on vertical grids of

▲ Most of the tower construction took place inside four enclosed boxes called "birdcages." Little construction activity was visible from outside the boxes, making the towers seem to magically rise. Inside, the birdcages were dens of noise, sweat and swinging hammers. Here, Clark Cottier, left, and Kelly Yarnell construct a portion of the dense web of reinforcing steel that rose from the caissons to the tower tops.

▲ A crane hoists the first birdcage into place on the tower pedestal on the Gig Harbor piertop.

rebar and beating on stubborn steel with sledge hammers. They worked as if possessed, shouting good-natured insults at one another and bellowing along with the rock music blasting from radios. The music was cranked up full volume but usually was barely audible in the racket of steel on steel.

They seemed to be having a great time, and Kelly Yarnell, a 34-year-old carpenter from Lake Tapps, Washington, said that was indeed the case.

"I love it," Yarnell said, pausing for a few minutes in the din of the birdcage on the Tacoma tower. "In our trade it's like a once-in-a-lifetime job."

Yarnell, who had biceps the size of most people's thighs and wore a diamond stud in his left earlobe, had worked in steel and concrete construction for 14 years. He had just finished helping build the bridge caissons, which gave him perspective for comparison.

Compared to the caissons, he said, working in the birdcages was a relief. "You're out of the elements," he said, "and there's a slight breeze that comes through here, which is nice."

Yarnell and other tower crew members kept two shifts going, from 6 a.m. to midnight, every day but Sunday. Their daily commute included a quarter-mile hike along one of the construction walkways suspended under the old bridge and then a 210-step descent to the tower base. There, they crossed over to the top of the new caisson on a gangplank and took the elevator up into the birdcage. At the end of each day, they made the same trip in reverse.

In all, Yarnell and the other crew members installed 1.25 million pounds of steel reinforcing bar in each tower leg. The vertical steel began at the tops of the caissons, where bars as big as baseball bats were embedded 12.5 feet into the concrete caps. As the towers grew upward, the workers threaded

extensions onto the 192 vertical bars so that each effectively became a single continuous piece of steel running from bottom to top. Into them they wove horizontal steel bars in a pattern so dense it was difficult to see though.

To avoid corrosion, all the steel had to be covered by at least 1.5 inches of concrete. Quality control inspectors were constantly on the prowl with rulers, measuring the distance from rebar to the inside surface of the forms. If the distance was less than 1.5 inches, the rebar had to be resituated.

▶ Once the concrete started flowing, it could not be interrupted until the entire section was finished. Pours often stretched late into the night.

◀ Cranes lifted wet concrete to the birdcages in four-yard buckets. Here, workers open the bottom of a bucket and release concrete into forms.

The most challenging aspect of the job, Yarnell said, was the lack of space. The walkways that ran along the perimeters of the birdcages were so narrow in places that the workers, most of whom were built with the big shoulders and upper arms of wrestlers, had to turn sideways to pass through.

When the forms and rebar were in place, concrete workers guided the wet mix into place and jostled it with vibrators to eliminate air pockets. The concrete was a special mix, designed to flow easily but still be strong and dense enough to keep salt water and acid rain from penetrating to the steel.

To reduce the air spaces in the concrete and increase its strength, Glacier Northwest added extremely fine particles to the cement used to bind the mix together. They used fly ash and silica, manufacturing byproducts so fine they are typically thought of as air pollution. In relative terms, if a grain of typical construction cement were the size of a basketball, a grain of fly ash would be the size of a softball. A grain of silica would be the size of a golf ball. When fully cured, the concrete would withstand a pressure of 10,000 pounds per square inch without breaking or crumbling. That was about twice the strength of typical structural concrete.

Work on the towers began in July 2004 and continued through the fall and into winter with the birdcages steadily ratcheting themselves higher and higher above the Narrows. While workers in one birdcage built forms and tied rebar, those in the other leg poured concrete and removed forms.

As the towers grew, workers had to adjust to the extreme height. Seagulls flying below them were mere specks. They also had to adjust to the disorienting tilt of the boxes. Because the tower legs tipped slightly toward each other, the walls of the birdcages were out of plumb.

▲ The tower legs were connected by cross braces, or "struts," which presented serious construction challenges in themselves. Here, workers building walls for the middle strut on the Gig Harbor tower tie rebar as the birdcages rise above them.

▲ Like the pistons of a slow moving machine, the four birdcages seesawed back and forth on their way to the top. Crews crossed from one birdcage to the other on high-altitude gangplanks. As workers poured concrete in one birdcage, another crew assembled rebar in the other.

Adjusting to the slant was surprisingly easy, said Bill Norris, the day-shift carpenters' foreman on the Tacoma tower. "You just start working, and it just becomes old hat," he said. "You don't ever pay any attention to it."

Winter weather required a more difficult adjustment. By November the birdcages had crawled hundreds of feet in the air and were exposed to the Narrows legendary wind, which frequently carried frozen rain and snow.

"With the cold and the wind velocity out there, it's going to bother anybody," said Dane Marbut, a state quality-control technician whose job kept him on the towers most of the time. Like Yarnell, Marbut had worked through the previous winter on the caissons. The towers were much colder, he said, because they were up off the water and exposed to the wind.

He dressed in layers and added waterproof boots, a knitted hard-hat liner and neoprene gloves. "Sometimes you're wet and cold and tired and want to go home," he admitted. "That's when you just have to think of a happy place and get on with the task at hand."

Earl White, a retired ironworker who helped build the 1950 bridge, was not impressed by stories of how cold it was up on the new towers. The birdcages provided workers with shelter from the wind, he said, something their predecessors didn't have.

"They're practically working in houses up there," he said. "We didn't have that luxury. I think they got armchairs inside those things."

White spent two winters working on the Narrows and said he was never more miserable. "I worked on towers in Cut Bank, Montana, and Havre, Montana, in the middle of winter, and I never suffered as much as I did on the Narrows. It can be a bearcat out there."

▲ Bird's eye view of new and old Gig Harbor towers, shot from the boom of the tower crane.

▲ The birdcages were 500 feet from the water at their final positions. Even at that height, the tower cranes soared above them. This is the crane operator's view, looking into the top of the birdcage in its final position on the south leg of the Gig Harbor tower. Workers called the top sections "wedding cakes" because they needed to incorporate so many complicated fittings for the saddles and the spinning operation.

"I don't know," the man said. "The gauge only goes up to a hundred miles an hour, and it went over that."

"That was the longest night of my life," White said.

THE CROWNING TOUCH

The birdcages made their way above the 500-foot level in June 2005. At that point, the tower cranes plucked them off, leaving the raw structures standing on their own. That marked the finish of tower construction, but the crowning touch was still to come: the installation of the saddles, the four 36-ton chunks of steel that would cap each tower leg and cradle the main suspension cables.

The saddles made suitably grand crowns. They were forged of nickel, chromium and molybdenum alloy steel and machined to tolerances measured in thousandths of inches. They were also strong, capable of withstanding the tremendous force the cables would put on them—about 60,000 pounds per square inch—without changing shape.

In a satisfying cycle of reuse, the Atlas foundry melted down many of the 240-pound links of anchor chain that had held the caissons in place and used them as part of the saddles' raw material.

If the tower saddles had been cast in single pieces they would have been too heavy for the tower cranes to lift, so Atlas made each one in two 18-ton pieces, which were to be hoisted separately and then bolted together.

The installation of the saddles was a significant symbolic event in bridge construction, signaling the completion of the towers and the transition to the cable work that would follow. But the event had added significance at the Narrows because of a

White remembered storms that coated the catwalks between towers with solid ice. One night, he said, the wind blew so hard he took shelter in a makeshift shack suspended between the catwalks. The wind bounced the shack up and down like a yo-yo and then started tearing metal sheets off its sides.

The tower had an anemometer mounted on top, White said, and he called up to another worker there to see what the wind speed was.

"How damn fast is it blowing?" White asked.

◀ Retired ironworker Earl White, who helped build the 1950 bridge, holds a photo of his old crew. White is standing in the back. To his right is Whitey Davis, who fell to his death.

▶ Workers guide a 20-ton saddle into place at the Tacoma anchorage. Saddles at the anchorages and tower tops were designed to cradle the wire in the main cables.

freakish accident that occurred in 1949. The first saddle was placed on the Tacoma tower just days before the worst earthquake in the South Sound's recorded history. The April 13, 1949, quake, centered just a few miles south of the bridge, rated 7.1 on the Richter scale, enough to set the bridge towers swaying back and forth like treetops in the wind.

"When that earthquake hit, the saddle wasn't fastened down in any way. It was just sitting up there on top of the tower," remembered Bill Matheny, a retired ironworker who helped build the bridge. "As the tower swayed back and forth, that saddle started moving."

Phil Orlando, an ironworker and local tavern owner, was alone on top of the tower when the earthquake hit. He had been stationed on the narrow platform, 507 feet above the water, to guide the crane operator.

The 28-ton saddle slid across the tower top toward Orlando, forcing him to scramble over it to avoid being swept off the tower. When the swaying tower reversed direction, the saddle paused, then headed back toward Orlando.

"He had to go over the top of it again," said White, who also remembered the incident. "He was a big man, but when you're scared you forget about the weight."

After a few more back and forths, the centrifugal force launched the saddle off the tower. It rocketed downward and scored a direct hit on a work barge anchored below. The saddle passed through the barge and kept on going.

After searching for three days, divers found the saddle partially buried on the bottom of the

> "He went down and got his paycheck, and he never came back."
>
> —EARL WHITE

Narrows, 140 feet beneath the surface. They fastened a line to it and a crane hauled it to the surface. The saddle was not significantly damaged and ended up back on top of the Tacoma tower.

Orlando was badly shaken by his close call, White remembered. "As soon as the earthquake was over, he went down and got his paycheck, and he never came back."

There was no such drama in 2005.

The saddles settled neatly on the tops of the towers in late June and workers bolted them down without incident, observed by small groups of bridge-project employees and journalists gazing over from parallel perches on top of the old bridge.

The newly hatched towers looked raw at first, with their struts still under construction, temporary wooden bracing still in place and work elevators still attached. Their greater size and plain grayness made the old bridge suddenly seem ornate and delicate in comparison. Even in its undressed state, the newcomer's character was clear. It would be the old bridge's square younger brother, solid and dependable but without much of a sense of humor.

Some would argue the point, but in Tim Moore's view, there is beauty in pure functionality, in the knowledge that every detail is in place for a purpose and contributing to a balanced system, spare and efficient. "Just the fact that it's a suspension bridge makes it beautiful," Moore said. "You don't have to gimmick it up."

And, he said, "Let's face it. If we wanted to add a lot of architectural detail, it would add a lot of cost."

▲ When the birdcages and forms were stripped away, the character of the new concrete towers was revealed: clean and unadorned, with function trumping form in nearly every case. The bas-relief X's in the struts of the new bridge, designed to echo the crossed steel braces on the 1950 towers, were among the few purely decorative touches.

SPINNING INTO TROUBLE

Nineteen thousand miles of wire, a celebrated Scotsman and a devastating outbreak of white rust.

As the towers neared completion, the area around the Tacoma end of the bridge began looking as if a carnival might be coming to town. A collection of odd-looking contraptions and whirligigs appeared on the blacktop near the anchorage: large spoked wheels, red and green spinning drums and a steel tower that looked like the makings of a Mad Mouse ride.

The elaborate equipment would be used to build the bridge's two main suspension cables, employing a process that sounded as improbable as the apparatus looked. Wire about the diameter of a drinking straw would be looped back and forth from shore to shore and across the tower tops 17,632 times, then compacted into two 20.5-inch diameter cylinders, each more than a mile long.

The wire was waiting nearby, stacked in bundles along the verge of Highway 16. It was hard-tempered, galvanized steel, made at a

▲ **Once the towers were completed, the next step was laying out construction walkways, or catwalks, from shore to shore. Chad "Hippy" Lyon, left, and Anthony Ray unfurled the wire-mesh walkways from reels on the tower tops, using only cables for support. The two 12-foot-wide catwalks, each more than a mile long, were used as temporary platforms to spin more than 19,000 miles of wire into the main cables.**

▲ **TNC engineer T.J. Paul inspects one of hundreds of bundles spinning wire, shipped from South Korea and stockpiled near the Tacoma anchorage.**

Kiswire manufacturing plant in South Korea and wrapped in bright yellow plastic. Altogether there was 19,115 miles of it, enough to wrap twice around the continental United States.

The cable-spinning apparatus was not designed by Rube Goldberg, but it looked as if it might have been. When assembled, it would combine aspects of sewing machines, cable cars and fishing reels. To accommodate all the necessary fittings and machinery, the top sections of the towers were so complex ironworkers called them "wedding cakes."

The cables for suspension bridges are assembled in this incremental way because they need to be so big. To be strong enough to hold up the 53 million-pound weight of the new Tacoma Narrows bridge deck, each cable would need to contain 5.2 million pounds of steel, far too much to consider building the cables on the ground and lifting them into place.

Aerial spinning was fairly common in the United States during the 1930s and 1940s, the golden days of American suspension bridge building. But by the turn of the 21st century, it had become a rarity. The spinning operation on the San Francisco Bay Area's Carquinez bridge in 2002 was the first of significant size in America since New York's Verrazano-Narrows Bridge connected Brooklyn and Staten Island in 1964.

Aerial spinning is a specialized skill, one learned largely through on-the-job training and requiring an array of expensive equipment. For an entire generation of American construction engineers and workers, it was a lost art. Neither Bechtel nor Kiewit, Tacoma Narrows Constructors' parent companies, had ever built a suspension bridge before.

They found their expertise in Dave Climie, an easygoing Scot whose accomplishments had put him among the world's top superstructure engineers. When Bechtel headhunters approached Climie for the Narrows job in 1999, he was only 38 years old but already had masterminded the cable and deck systems on three of the five largest suspension bridges in the world.

Despite his vaunted stature among bridge engineers, Climie was a modest, unassuming man who brought infectious enthusiasm and a refreshing lack of pretension to the TNC organization. His profession kept him traveling around the world and cost him a stable home life. To have to transport a wife and family every time he changed projects did not seem reasonable to him. "I'd be on my third divorce by now," he said.

▶ **Saburo Matsueda, a suspension bridge specialist with the Japanese subcontractor, NSKB, helps haul down the 5/8-inch steel cable that became the first connection from shore to tower.**

▼ **Cable spinning required specialized equipment not commonly used in the United States in 40 years. As spinning begins, Danish engineer Karsten Baltzer, left, and T. J. Johnson adjust a wire organizer.**

Climie brought another cable specialist with him, an engineer named Karsten Baltzer, the son of a draftsman from the town of Sorø, in Denmark. After engineering school, Baltzer had the good fortune of speaking Danish when the Storebaelt East Bridge was being built across Denmark's Great Belt, giving him a competitive edge in landing a job there. He worked with Climie in Denmark and also in China on the Jingyiang Bridge across the Yangtze. Immediately before coming to the Narrows, Baltzer had managed the spinning operation on the Carquinez bridge.

Climie and Baltzer represented the European school of suspension bridge building. Another more active contingent was making innovations on the other side of the world.

Toward the end of the 20th century, the cable-spinning expertise that once had been primarily the province of America and Northern Europe moved to Japan. In 1998, after completion of the world's longest suspension bridge, the Akashi Kaikyo, Japanese firms began aggressively marketing their suspension bridge building expertise overseas. In the 18 months after the Akashi Kaikyo Bridge opened, marketers from Nippon Steel visited every country in world where new bridges were under some stage of construction.

Nippon's marketing efforts won them contracts on the Kwang-Ahn Bridge in Pusan, South Korea, which opened in 2002, and the Runyang Bridge on the Yangtze in China, which became the world's third largest suspension bridge when it opened in April 2005. In 1999, when Bechtel and Kiewit were considering subcontractors that could handle the cable production and spinning operations at the Narrows, a joint venture of Nippon and Kawada Bridge, made a persuasive case for themselves.

Kawada had been building bridges in Japan since 1927, and, in cooperation with Nippon on the Akashi Kaikyo, they had built cables 44 inches in diameter and containing 50,000 tons of steel, many times larger than those required for the Narrows.

In 2002, the Nippon-Kawada partnership, doing business as NSKB, jointly won the contract to manufacture and erect the girders and cables for the new Narrows bridge and to lease spinning equipment.

Nippon and Kawada sent not only shiploads of spinning equipment and wire to the Narrows, but also six suspension bridge specialists, all of whom previously had worked on at least other four suspension bridges.

The first members of the Japanese team arrived with their families in March 2005 for an assignment expected to take one year. All moved into the same apartment complex, the Cliffside, on a bluff overlooking the Narrows, just a few minutes walk from the bridge construction site and TNC's Narrowsgate field office.

The team included two design engineers, Kiyotaka Miwa and Shin Fukatsu; construction engineer Yoshinari Takai; and mechanical engineer, Takenori Watanabe, who would supervise use of the spinning equipment.

Katsuhide Mine and Saburo Matsueda, field engineers assigned to instruct American workers, had a solid knowledge of cable spinning, but a shaky grasp of English. They demonstrated spinning techniques to ironworkers with an expressive array of sign language and body English.

THE FIRST CONNECTION

As with spider webs, extending the first filament across space is a critical and exhilarating step in the

construction of a suspension bridge. At the Narrows, workers had spent three years establishing supports for the cables: the anchorages that would hold them in place on shore and the towers that would support them over the water.

Now they were ready for the step that would take them from earth to air, the realm that defines the medium.

On July 21, 2005, winch operators on top of the Gig Harbor tower lowered the end of a 5/8-inch cable to deck hands aboard the tugboat Henry Foss, idling below. Meanwhile, on shore, workers dragged another cable down the steep bluff from the Gig Harbor anchorage to the beach and handed it over to workers on an aluminum work skiff.

The tug and skiff met, deck hands clipped the ends of the cables into a triangular steel connector plate, and, shortly before noon, the winch operator on the tower top began reeling in his end. As the cable tightened, it sliced through the water, then sprang into the air, dripping water and glinting in the sunlight. On one end was the anchorage, 81 million pounds of concrete buried in the earth. On the other end, a quarter-mile away and 50 stories in the air, was the top of the tower.

The occasion received no official recognition, but bridge engineers and workers celebrated informally. Linea Laird, the state's project manager, hiked a half-mile along the cobbled beach from Narrows Park to get a close-up view of the operation. When the connected cables rose into the air, Laird stepped back and snapped a photo of a handful of her top managers, who also had come down to watch.

"It's our first line," Laird said. "You've got to get that first cable across before you can get all the rest."

Over the next several days, TNC crews, guided by their Japanese mentors, used similar operations to link the Tacoma tower to the Tacoma shore and then complete the crossing by linking the two towers across the main span. They used the first cables to ferry others, clipping them on with hangers and hauling them aloft.

Knowing the weight of the cables and deck would pull the towers toward each other like trees bent by a heavy hammock, TNC prepared by tilting the towers toward the shore. Using eight heavy cables attached to the anchorages, TNC hauled back on the tower tops, pulling them nearly 2 feet out of plumb. Miwa, NSKB's chief design engineer, had

▲ A TNC worker clips the cable ends into a triangular connector and signals they are ready to hoist aloft, establishing the first shore-to-tower connection.

▲ Tugboats carrying the ends of cables from both tower tops meet in the middle of the Narrows to link them together.

calculated that when the main cables were finished and the deck attached, their combined weight would pull the towers back to a straight-up-and-down position.

Sixteen more cables were hauled up to support hanging construction walkways workers would use during the cable-spinning process. Two parallel walkways, or catwalks, each 12 feet wide and more than a mile long, would reach from anchorage to anchorage, tracking the eventual paths of the main cables across the tower tops.

The catwalks were made of welded wire mesh that resembled heavy-duty hog-wire fencing. To keep workers from slipping as they walked, perpendicular wooden slats were wired onto the mesh, 1 foot apart on the steepest sections and 2 feet apart elsewhere. Four-foot-high mesh walls were clipped onto the sides and supported by another set of cables.

The catwalks were temporary, but they needed to be carefully designed to withstand the Narrows' legendary winds. Crosswalks between the catwalks are standard stiffening features on suspension bridge construction, but at the Narrows, designers called for twice as many as usual—two on the side spans and five in the midspan—to keep them from flying around or blowing into the existing bridge.

"This is our major working platform for all of the suspension system," Climie said. "It will be up there for the best part of a year and a half. When we're spinning the main cable we'll have 40 or 50 people strung out working on it, so it has to be built to keep them safe."

Long sections of the catwalks were assembled on the ground and then hoisted to the tower tops in cumbersome rolls. There, workers hooked them onto their support cables and gradually unfurled them, letting their weight carry them down the slopes.

The cables hauled aloft to tilt the towers and carry the catwalks were so slender they were barely visible to drivers on the old bridge. But when workers unfurled the first section of catwalk from the Gig Harbor tower top, the effect was startling. Thousands of commuters watched as the stub of walkway reached outward, seemingly magically suspended in midair.

The illusion was soon dispelled. On August 10, 2005, at the height of the afternoon commute, the first 160-foot section of catwalk got away from workers at the tower top and slid a few hundred feet down the cables, moving like a runaway roller coaster. Restraining cables abruptly yanked the section to a halt midway down, collapsing it into a midair tangle of wire mesh, cables and boards.

No one was hurt, but the mistake, hanging in plain view, was painfully embarrassing to Manuel Rondón, who took great pains to assure that TNC was never seen in an unfavorable light. After a day of analyzing what had gone wrong, workers unsnarled the mess and continued building. Engineers pronounced the catwalk section undamaged. When straightened, it was indistinguishable from the rest.

The catwalks on the new bridge were 4 feet wider, but otherwise remarkably similar to those used in the construction of the two previous Narrows bridges, in 1940 and 1949. When those catwalks were finished, walking on them was a perk freely handed out to politicians and friends and relatives of bridge officials.

Dozens of people made the spectacular, high-altitude walk in 1940, including the wife of the state's chief bridge inspector, Harvey Donnelly, and Marie Guske, construction engineer Clark Eldridge's secretary and the only woman employed on the bridge job. The two women drew special attention for making the mile-long traverse in high heels.

▲ Ironworker Russell Scott, on the Gig Harbor tower top, tugs on the last few feet of a roll of catwalk wire, helping gravity carry it down supporting cables.

Such thrills would be in short supply this time. Rondón was not about to compromise safety by allowing sightseers on the job site, and he felt no need to curry favor with local politicians. The entrances to the catwalks were protected by locked chain-link gates. With few exceptions, only the workers who needed access got inside. Anyone who set foot on the catwalks was required to take a safety class and wear a webbed safety harness.

Working on the catwalks took some getting used to. Their upward slope began gradually at the anchorages, but the closer they got to the towers, the steeper they became. Near the tops they rose at a 25 percent grade. Climbing on them there was a heart-pounding effort, made thrilling by the height

---

◄ PREVIOUS PAGE: Workers unrolled the catwalks from rolls hoisted to the tops of the towers and slid them along supporting cables. Six cables, each about the diameter of a tennis racquet handle, supported the catwalk floor. Two slightly smaller cables supported 4-foot-tall mesh side walls woven into the floors. When finished, the catwalks were nearly impossible to fall from, but they still caused some psychological stress. To improve footing, TNC wired wooden slats onto the mesh, 1 foot apart on the steepest sections and 2 feet apart elsewhere.

▼ Nearly 500 feet over the Narrows, an ironworker helps the "sled" carrying the leading edge of the catwalk slide down supporting cables.

► To keep the parallel catwalks from swinging too much in the wind, TNC connected them with 4-foot-wide cross-walks, two on each side span and five in the center span.

Five electricians, subcontractors from the local Tacoma company Totem Electric, were the first workers to try out the new catwalk, having been assigned to install fluorescent lights and power outlets for tools and equipment.

After their eight-hour shift, the electricians came off the catwalks weaving like sailors just off a ship. Foreman David Gaddy said the reality of the altitude became clear to him when his cell phone slipped off his belt. The phone fell neatly through the wire mesh floor and grew smaller and smaller as it floated through the hundreds of feet of empty space. Gaddy watched it fall, looking between the toes of his boots.

"It's a long way down," he said. "I counted five seconds before it hit the water."

What catwalk workers found the hardest to get used to, though, was not the height but the lack of stationary visual references. Everything was in motion, the clouds in the sky, the water below and the catwalks themselves, making it easy for workers to lose their bearings and start feeling queasy.

"If you're looking up and see the clouds moving one direction and a bird flying the other," Gaddy said, "it can really throw you off."

The trick was to stay focused on the work at hand, said Doug Marshall, a 48-year-old electrician from Lacey, Washington. "Everything's moving if you look too deep."

Another thing to get used to, Marshall said, was the difference in springiness between the outer edges of the catwalk and the middle. The edges were stiffened by boards, but the middle was not.

The difference surprised Marshall when he first ventured onto the catwalk from the top of the Gig Harbor tower. When he stepped off the boards and sank into the unsupported wire, he said, the sense of falling brought a rush of panic.

▲ Electricians stringing work lights were among the first to try out the new catwalks. The trick to avoiding vertigo was to stay focused on the task close at hand, said electrician Doug Marshall. "Everything's moving if you look too deep," he said.

▶ Fluorescent work lights, installed every 40 feet along the two miles of catwalks, produced a dramatic light show from the air.

and view. Seattle's Space Needle and high-rise office buildings, 25 miles away, were clearly visible, as was the entire Olympic Range and much of the Cascades. Seals swimming in the water below were so far down they looked like flagellated protozoa on microscope slides.

Even for people not bothered by heights, the catwalks were disconcerting because their open mesh floors and sides allowed a clear view to the water and sky below. Although falling out of them was almost physically impossible, the wire seemed to offer no protection. Making matters worse, every step produced a slight rebound, like walking on a trampoline.

apparatus. First, tramway support frames went up along the catwalks and at the tops of the towers, then the heavy haul ropes that would make endless circles from shore to shore, carrying spinning wheels.

The plan was to build both main cables at the same time, alternating between laying down wire and making the precise adjustments necessary to keep them hanging in the right profiles and at the right degree of tension. At the center of the midspan, the finished cables had to hang within a half-inch of what engineers had determined to be their ideal position.

When 464 wires had been laid down, they would be gathered together into a bundle like a big handful of uncooked spaghetti. Each finished cable would contain 19 of those bundles, made up of a total of 8,816 individual steel wires. The bundles would then be squeezed tightly together with hydraulic pincers and fastened with stainless steel bands. The spinning process, Climie estimated—optimistically, as it turned out—would take three months.

Twenty-eight workers were stationed on the bridge during spinning, spread out along the catwalks, on the tower tops and at the anchorages. In the production area behind the Tacoma anchorage, another half-dozen workers carried the four-mile coils of wire from the stockpile and loaded them into a machine that rewound them onto four larger drums set up on the anchorage. Each 7.5-foot drum took six coils of wire, or about 24 miles.

Wire from the four drums was looped over a spinning wheel that was pulled by the haul line toward Gig Harbor. As the wheel headed off across the Narrows, it drew wire off the drums like a trout running line out of a fishing reel. Workers positioned

"It gives you a pause to think about life," he said.

SPINNING

Once the catwalks were in place, the Japanese engineers guided the assembly of the spinning

▲ An aerial tramway similar to those at ski resorts carried spinning wheels over the towers. Here, a bargeload of tramway support frames waits to be hoisted onto the catwalks.

▲ With the catwalks in place, workers erected support frames for the tramways that would carry spinning wheels. Workers with backpacks make their daily commute at the beginning of their shift.

along the catwalks made sure the wires landed in their proper positions by placing them in organizers that looked like upturned pitchforks.

Most of the action was at the anchorages. There, the wheel would pause, giving workers time to pull the wires slack and hand the looped ends down into the anchorage where two more workers would "dog off" the wire—that is, hook it over semicircular "strand shoes" encased in concrete. Then they'd flip the wheel into position to draw more wire off the supply drums and start over again. And again, and again.

The spinning process had changed little since John Roebling patented it in 19th century. But it was new to workers at the Narrows. The extreme height, the big spoked wheels speeding overhead like riderless unicycles, the moving cables, all made the job seem complex and dangerous. At top speed,

---

▼ To resist the 25 million pounds of force each main cable exerted, Dave Holcomb, left, and Richard Smith secured strands of spinning wire to steel rods embedded deep in concrete at the rear of the anchorages. By turning nuts on the 4.5-inch threaded rods, engineers could tighten or loosen cables, like tuning a mile-long musical instrument.

▲ The spinning wheel carries the first four strands of wire down from the top of the Tacoma tower as, from right, Tracy Martin, Joe Gonzales and Karsten Baltzer watch to make sure the system is operating as it should. The cowbell attached near the hub of the wheel is a suspension bridge-building tradition, intended to warn workers that the wheel was approaching. At top speed, the wheel moved about 14 miles per hour.

Up to that point, Rondón's witches behind the door had been amateurs, short-hitting adolescents just learning their trade. They had succeeded in getting Tom Sherman's attention on the caissons when the anchor chains broke but did little damage and caused only minor delays.

The witch that appeared November 12, 2005, was of another caliber altogether. That day, workers in the production area sliced open the yellow plastic covering one of the thousands of bundles of stockpiled wire and did a double take. The top coil was covered with fluffy white powder, as if infected with a bad case of dandruff.

They called supervisors, who, after a quick look, began striding from bundle to bundle, tearing wrappers open. They found white powder in another roll, then another, then dozens.

The white power was oxidized zinc from the wire's galvanized coating. The zinc was intended to oxidize, but not this way. Normally, exposure to air induces zinc to form a stable, nonporous form of zinc carbonate that makes a hard, protective coat for the steel beneath.

Instead, this zinc reacted differently, forming a porous form of zinc carbonate sometimes referred to as "white rust." It ate the zinc away, leaving the steel exposed to the weather and prone to oxidize and weaken. The damaged wire was useless for spinning.

Inspectors sorted through the stockpiled wire during the week of Thanksgiving, separating good from bad. They discovered that 600 of the 3,500 coils remaining—or about 2,400 miles of wire—were so badly corroded they were unusable. Another 400 coils were questionable.

At a cold news conference near the Tacoma anchorage on November 30, Linea Laird, shivering in a canvas jacket too thin for the day, announced to

as to automatically program the tram with the spinning wheel to speed up on the downhill runs, slow at the tower tops and pause inside each anchorage so workers could dog off the wire and reposition the wheel for the next trip.

Each one-way trip took between 11 and 12 minutes, plus three or four minutes to hook up the wire in the anchorage. For some, that left time for magazine reading and admiring the scenery.

One afternoon, Andy Hoff, one of TNC's top engineers, looked as if he might fall asleep as he watched spinning crews at the Tacoma anchorage reverse the wire for the next phase to begin. "I'm ready for the hundredth-odd time," he said.

▲ PAGE 106: As cable-spinning stretched into the winter of 2005-2006, cold and wind made conditions barely tolerable. A tattered flag shows the effect of nearly constant wind.

▲ PAGE 107: With little to do between passes of the spinning wheel, workers often were cold and alone on the catwalk. Ironworker Jason Anderson takes cover in a plastic shelter.

a group of reporters that the bridge project had hit a serious snag. Behind her, workers on forklifts darted from the stockpile to inspection stations, carrying wire to a growing pile of rejects.

"We thought it must be an isolated instance," Laird said. "We had no idea of the magnitude."

NSKB and Kiswire, its South Korean subcontractor that manufactured the wire, was not to blame, Laird said. TNC and the state had quality control inspectors at the Korean plant, and they had verified that the wire was manufactured and coated correctly.

The problem, Laird said, was improper storage. Keeping the wire outside under plastic combined the factors most likely to produce unstable zinc oxides: a constantly moist environment and limited oxygen flow. With four television cameras running, Laird flatly said, "We improperly stored the wire." She implicated TNC and the state Department of Transportation equally, pointing to the logo on her jacket that linked the two: "One Project, One Team."

Rondón was dismayed when he heard what Laird had said. In his view, too little was known about what had happened for anyone to start accepting blame. Early analyses made it appear as if there might be some correlation between how the wire had been wrapped in South Korea and the amount of corrosion present.

Two types of wrapping had been used at the Kiswire plant. One was all plastic; the other was plastic with a paper lining. Coils wrapped with the plastic and paper seemed affected the worst, giving rise to the notion that the paper might have retained moisture, creating the conditions that caused the white rust.

Replacing thousands of miles of ruined wire would cost millions and possibly delay the opening

of the bridge. The question of blame very likely would wind up in court. Rondón was not ready to concede anything.

Since TNC would have to bear the cost of replacing the wire, the Department of Transportation's main concern was that bad wire might already have been spun into the bridge cables. That concern was put to rest by reviewing inspection records. Quality control inspectors from TNC and the state had routinely inspected wire as it was reeled and had kept samples of each coil for reference. No corrosion was evident on any of the wire.

Once that had been determined, the state's job was reassuring the public. They launched a public relations effort to assure people that no corroded wire had gotten into the cables.

"There is no bad wire on the bridge," Laird said. "It's a very visual thing. You can see it immediately if it's there."

Transportation Secretary Doug MacDonald weighed in, giving his assurance that only wire meeting contract specifications would be used.

"We expect the contractor to obtain additional wire, if necessary, to assure the timely and correct assembly of the suspension cables," he said.

Chemical labs hired by TNC and the state took samples to assess how much zinc coating remained on the wire, separating it into four categories ranging from "Unaffected" to "Inappropriate for Spinning." Workers cleaned wire that was only lightly corroded by pressure washing it and brushing with stainless steel brushes.

Outwardly, Rondón projected confidence. But back in the Narrowsgate office, he and Climie were scrambling to keep the bridge project on schedule. TNC's contract with the state had no specific deadline for cable spinning. It specified only that the first

► A close-up of "white rust," the harmful oxidization of the zinc coating on the steel wire. The unstable chemical reaction ate away the zinc, leaving the steel vulnerable to the weather.

► Linea Laird, state manager of the bridge project, announces a major setback at a cold news conference on November 30, 2005. More than one-quarter of the 19,000 miles of stockpiled spinning wire had corroded, rendering it unusable.

Climie and Rondón called emergency conferences with their top engineers, looking for brainstorms.

It was Karsten Baltzer, the Dane, who came up with the idea that saved the day: Rather than continuing to work on both cables at once and shutting down the entire spinning operation when the good wire ran out, he said, why not put every effort into finishing just one cable? That way, they could keep crews busy compacting and banding the finished cable rather than laying everybody off. When the new wire arrived, they would have experienced crews still on the job, ready to speed through the other cable knowing exactly what to do.

Rondón was delighted. He wanted to hug Baltzer, to dance around the room with him. As Baltzer suggested, the spinning crews shifted all effort to the south cable, keeping their fingers crossed that there would be enough good wire to finish.

Meanwhile, Rondón and Climie set off on a worldwide search for replacement wire. It was a difficult quest. The wire was highly specialized and factories were booked with other jobs. After a month they at last reached deals with three wire manufacturers. The Kiswire plant in South Korea could turn out a portion of what was needed. Plants in China and England would do the rest. Unfortunately, none of the wire could be delivered until March at the earliest.

In January, Rondón acknowledged for the first time that TNC might miss the state deadline that called for having the first of 46 deck sections in place. He emphasized, however, that the May date was an interim deadline only and that he still intended to have the bridge finished and open to traffic in April 2007.

It seemed possible. Meanwhile, however, other problems were brewing across the Pacific.

▶ To salvage wire that was only lightly corroded, workers blasted it with high-pressure water and brushed it with wire brushes.

▶ TNC had enough wire to finish one main cable but had to wait for replacements to finish the other. Here, James "Sully" Sullivan, who at 78 years old was the oldest worker on the bridge, hefts a pair of giant pincers used to compress the finished strands for temporary banding.

deck section must be lifted into place on the finished cables by May 7, 2006, less than six months away.

TNC's internal schedule had called for finishing the cables before the end of the year, a goal now clearly impossible. It looked as if, when the good wire ran out, they would have to shut down the spinning operation, lay off workers and wait for more wire to be manufactured and delivered. By that time, the workers almost certainly would have other jobs, forcing TNC to waste more time and money retraining new crews.

THE SAMSUNG SAGA

Building the deck turns out to be more complicated than TNC's Asian partners bargained for. Demands for more money lead to a work stoppage, lawsuits and missed deadlines.

Dave Climie was looking for the perfect bulgogi restaurant. It was late and businesses were closing for the night, but Climie was convinced the right place was still out there. He strode through Pusan's crowded Gwangali district like a hound on the scent, dodging motor scooters and taxis and threading through knots of people crowding intersections. One restaurant after another failed to meet his high standards.

Perfect, in Climie's mind, meant a barbecue grill in the middle of the table, plenty of cold beer, and heaps of marinated steak and vegetables to cook over the hot coals. And it meant having a view of the Kwang Ahn suspension bridge, South Korea's longest.

After searching one crooked alley after another, the hunt started seeming hopeless. But at last, there it was: tables on a deck out front, grills glowing, and a scent of soy sauce, steak and garlic in the air. A waiter smiled and beckoned toward an empty table with a panoramic view of the Kwang Ahn bridge, blinking and shimmering on Suyong Bay.

"What did I tell you?" Climie said, settling in.

Climie is an unmitigated optimist. Whatever the problems, he tends to believe they will work themselves out and that order will prevail.

It's an attitude that has served him well in bridge construction. One of his responsibilities as superstructure manager of the new Tacoma Narrows bridge was seeing that the Korean ship-

▲ On Koje Island in South Korea, Samsung workers check the inventory of Japanese steel they will cut and weld into 46 deck sections. Using Japanese and Korean subcontractors to build the deck saved TNC millions in labor and raw material costs.

builder, Samsung Heavy Industries Inc., finished the new mile-long deck on time. The process tested even Climie's patience. In contrast to the new bridge's caissons, anchorages and towers, all of which went smoothly, the deck was plagued with problems and delays from the beginning.

As Climie settled in at the restaurant, the 35.5 million-pound steel structure was under construction at the Samsung shipyard on Koje Island, 20 miles off Korea's southern coast. It was months behind schedule and already had cost millions more than it was supposed to. Lawyers were churning out briefs in preparation for lawsuits.

Tacoma Narrows Constructors' deadline for having the first deck section hoisted into place was just 11 months away, and only four of 46 deck sections were finished. Climie had shuttled back and forth across the Pacific so many times he had used up all the pages in his passport and had to get a new one.

Climie was used to a certain amount of this kind of angst. Deck-building is almost always the last act in bridge construction dramas. At that point, deadlines are looming and budgets are almost always strained.

"By the time they get to the superstructure they run out of time and money," he said, grinning.

Building the deck out of Japanese steel and fabricating it in South Korea was to have been a huge money saver for TNC. Japanese steel was cheaper than American, and labor costs in South Korea were about 30 percent lower than in the United States. Samsung's expertise in fabricating steel ships and offshore drilling rigs seemed to make it ideally suited to build the complex bridge framework. During contract price negotiations, a project cost analysis showed using foreign steel and foreign labor would save the Narrows project about $30 million.

But the Asian partnerships came with problems.

"We have three different countries, three different cultures and three different ways of doing things," Climie said. "Sometimes that leads to misunderstandings."

As it turned out, the predictable misunderstandings attributable to culture, language and distance were minor compared to the problems that arose from the nature of the contract TNC signed with Washington state. Because it was a "design-build" contract, it required that final prices be established before the design was finished. That meant uncertainties not only for TNC but for its

▲ Samsung had built many bridge decks but never one so complex. Its main business on Koje Island was building ships. The shipyard was the third-biggest in the world—so large, bridge workers used bicycles to travel between work stations.

▼ Dave Climie, left, and Paul Lange, a TNC welding expert, in TNC's cramped Koje Island office, where they monitored Samsung's work.

subcontractors as well. They had to come up with dollar amounts before they knew exactly what they were getting into.

The Japanese industrial giants, Nippon Steel and Kawada Bridge—doing business in a partnership called NSKB—won the contract to supply the deck. NSKB, in turn, subcontracted the fabrication work to Samsung.

Samsung bid on the work in 2002, basing its estimates on design drawings that at the time were only 10 percent complete. When Samsung received TNC's finished design in September 2003, it saw that the job was going to take a lot longer and cost a lot more than it had expected.

After eight months of work, Samsung estimated that its total cost would be $79 million, more than three times the $24.6 million it was being paid. Samsung and NSKB complained to TNC, claiming its designers had made "radical changes" to the deck design after the bids were in. TNC maintained there had been no radical changes, just poor cost estimating by the subcontractors.

In December 2004, with the discussion heated and going nowhere, Climie and Manuel Rondón met with managers from Samsung and NSKB in Maui, Hawaii, for a three-day mediation session. The mediation was led by Antonio Piazza, an $8,000-a-day arbitrator highly regarded for his successes in breaking up construction industry logjams. Piazza based his techniques on aikido, the Japanese martial art intended to resolve conflict without escalating violence.

As a result of the Maui meetings, Samsung got another $12 million but still was not satisfied. A second five-day meeting with Piazza was scheduled but never took place because the sides were too far apart to talk.

A DIFFICULT DESIGN

The design of the new bridge deck at the Tacoma Narrows was of great interest in the international engineering community because of Galloping Gertie's notoriety. Leon Moisseiff's lightweight deck had proved spectacularly inadequate, and the film of its failure was imprinted in the minds of civil engineers from Australia to Ankara.

The deck that replaced Gertie's in 1950, also much watched, was not an elegant solution but a fail-safe effort, driven by determination not to let a similar disaster happen again. The 1950 deck was a 33-foot-tall strongman, so heavily reinforced with trusses it amounted to overkill.

TNC's new deck would be a different story. It was to be a model of strength and simplicity, incorporating engineering advances that would make it strong, light and resistant to corrosion. According to Tom Spoth and the Parsons design team he led, the clear choice for the job was an orthotropic deck.

"Orthotropic" is engineering shorthand for "orthogonally anisotropic," which means structures whose supporting components lie at right angles to each other. An orthotropic deck bridge is put together like the floor of a house, with joists running one way and supporting girders running perpendicular beneath them. The tops of orthotropic bridge decks are made of relatively thin sheets of solid steel, which not only form the base for the driving surface but also contribute to the overall rigidity of the structure.

Orthotropic decks were first used after World War II in Germany, where hundreds of bridges had been destroyed by Allied bombers and needed to be replaced quickly and efficiently. Thanks in large part to new welding technology, the new bridges were spare and efficient, far lighter than old-style bridges

▲ Under the watchful eye of quality control officer I. K. Lee, far left, Samsung welders preheat and weld the understructure of a deck section. The sections, as big as apartment buildings, were constructed upside down, then flipped over with giant cranes.

▲ Fitting the deck sections together required measurements so precise that tolerances were measured in millimeters. Here a surveyor on top of one section checks vertical alignment as three of the sections are fit together in a dry run. (He's sighting toward his assistant shown on the following page.)

with their heavy underlying trusses and layers of reinforced concrete on their roadways.

Even though the new Narrows bridge would be 12 feet wider than the 1950 bridge, its floor system would weigh 16 percent less—3,360 pounds per foot versus 4,220 pounds per foot.

Despite their acceptance elsewhere, orthotropic bridge decks were relatively rare in the United States when plans for the new Narrows bridge were being drawn up. There were only about 50 of them in the country, but they included some widely praised successes—the new Al Zampa Bridge over California's Carquinez Strait, for example, as well as portions of the San Francisco-Oakland Bay Bridge and San Diego's Coronado bridge.

Orthotropic bridge decks were not at all unusual in Asia, however. Japan alone had more than 250 of them, including the world's longest. But the design of the Narrows deck had features that made it unusually difficult to build. The Washington State

Department of Transportation wanted the new deck to be visually compatible with the one on the 1950 bridge, and it wanted the new bridge to be able to accommodate a second level at some point in the future, with a minimum of retrofitting.

"The existing bridge is more than 50 years old," said Tim Moore, the department's bridge adviser. "In 2050, it will be a hundred years old. It's in extremely good shape right now, but it's not going to last forever."

So, rather than supporting the new deck with a simple box-girder support system like most orthotropics, Spoth and his team came up with an open steel truss design, deep enough to be modified to make room for another roadway or light rail system.

Spoth and his team designed the bridge so that the bottom lateral bracing and cross frames could be removed and replaced with a second orthotropic deck below the first. The added weight could be

▲ As three finished deck sections are fit together, several others are under construction in the background. All of the deck sections were preassembled on Koje to attain the arched shape requested by designers.

supported by new suspenders hung from two additional main cables draped across the tower tops.

Samsung had built many orthotropic bridges, but never one supported by a trussed steel framework. For them, that was unexplored territory.

WELCOME TO KOJE

A year into the deck-building effort, Climie's visits to the Samsung shipyard began fitting into a routine. He would fly from Seattle-Tacoma International Airport to Pusan via Tokyo, stay overnight at the Hotel Paragon, close to the airport, then take an early morning hop out to Koje on Samsung's private company helicopter.

From the air, the chain of islands that includes Koje easily could be mistaken for Washington's San Juans. Koje is among the largest of the islands, a cushy swatch of green velvet in the blue sea.

Peaceful though it seems, Koje's recent history has been anything but. During the Korean War, South Korea and the United States set up a Koje

▲ Checking the fit: Top construction officials from Samsung and NSKB confer with TNC's Paul Lange on a questionable joint between sections. Lange tells them that a 1/16-of-an-inch difference in elevation along a 2-foot section of the joint must be improved.

▼ Work at Samsung was monitored by four layers of quality control, including inspectors from Washington state. Here, state Department of Transportation inspector Dave Harkema confers with a Samsung quality control officer about the sealed U-ribs that will support the steel top plate.

world. Daewoo's is the second biggest. When the shipyards moved in, Koje had a population of about 2,000 people, mostly rice farmers and fishermen. By 2000 its population was 100 times that.

Every day, between 12,000 and 15,000 people came to work at the Samsung yard, which turned out from 35 to 50 ocean vessels a year—nearly one a week when business was good.

The shipyard was so big it had its own fire department and police force. Construction cranes as tall as 20-story buildings gave the yard its own spiky skyline. A fleet of Samsung buses ran 38 routes into the surrounding countryside to pick up and drop off workers.

At the yard, Climie shared an office with Rudy van der Maat—a Sean Connery look-alike from Brisbane, Australia, who was TNC's expediter—and Paul Lange, a welding engineer responsible for TNC's quality control. Next door was NSKB's tiny office, and across the hall was a cubbyhole for Washington state's quality control inspector, Dave Harkema.

Communication among the companies was complicated by their contractual arrangements. Technically, Samsung was not working for TNC. It was working for NSKB. That meant Climie and his team were supposed to go through NSKB instead of dealing with Samsung directly. The dynamic was cumbersome and uncomfortable, even for Climie, who was charmed with an ability to put people at ease.

"I like to get all parties in the room at the same time," Climie said. "Otherwise, it's like a Chinese whispering game."

The TNC office at Samsung was small, but the Narrows deck construction project took up the largest share of land of any single project in the 141-acre shipyard. The fabrication shop, where raw sheets of steel were cut and welded into deck components,

▲ Samsung inspectors, at right and in the background, ride hydraulic lifts to check out the underside of a nearly completed deck section.

▶ Members of the Samsung's Tacoma Bridge team line up in neat ranks for early morning exercises and a pep talk to prepare for the day's work.

prison camp, where they kept thousands of Chinese and North Korean soldiers captured during the 1950 invasion. A bloody prison riot in 1952 ended with 69 dead. The prisoners captured the American camp commander and forced him to sign documents saying they were being mistreated, a propaganda coup for the Communists.

Now, Koje is essentially a big company town, or, more accurately, a two-company town. Not only does Samsung have its shipyard there, but so does Daewoo. Samsung's yard is the third biggest in the

contained more than two acres of floor space.

Inside, the noise was intense, even with ear plugs. The steady cacophony of motors, hissing air and grinders was punctuated by sudden explosive concussions of steel on steel, loud enough to make you want to drop to your knees and hold your head. Cranes on overhead runners raced back and forth with sheets of steel in their magnetic grabbers, sirens screaming as they went.

Because the Narrows job was behind schedule, signs urging greater speed and efficiency had been posted throughout the construction area. Outside, a banner as big as a billboard read, "120-Day Recovery Fight to Get Back on Track."

During shift changes, men with signs stood at street corners, smiling and waving at arriving laborers like American campaign workers on election eve. The signs said, "Tacoma Narrows Project: Let's Show Them What Korea Can Do."

COMPLEX CONSTRUCTION

Seeing the designs for the new deck on paper was one thing, but seeing it under construction gave a fuller appreciation of the difficulties involved. Because the finished bridge needed to arch in the middle to make room for marine traffic, each section had to be slightly different so it would fit into the overall curve. The deck surface also was designed to curve from side to side for drainage, which added another degree of complexity. Before the sections were delivered to Tacoma, they needed to be tested to make sure they fit together. The allowable gaps between deck sections, which aver-

> "Let's show them what Korea can do."
>
> —Signs at Samsung shipyard

aged 120 feet long and 450 tons, were measured in millimeters.

Viewed in cross section, the top of each deck section resembled an extremely large sheet of corrugated cardboard. The $5/8$-inch steel slab formed the outer surface and, beneath it, closely spaced U-shaped steel troughs, or "ribs," ran lengthwise. The ribs were supported at even intervals by heavier perpendicular or "orthotropic" braces. Beneath those was the triangular trusswork, 24 feet tall.

The welds that attached the U-ribs to the steel sheet on top needed special attention, Climie explained.

"They're under constant stress as cars go by," he said, shouting to be heard over the noise in the fabrication shop. "There's a load, then no load. A load, no load. This weld is absolutely critical."

Spoth's design team specified to Samsung that the gap between the U-ribs and the deck top could be no greater than one-fiftieth of an inch. Also, the U-ribs were to be completely sealed to the weather.

That aspect of the design was intended to reduce corrosion and save the state millions in maintenance costs, but incorporating it took remarkable care. In all, on the 5,440 linear feet of deck, there would need to be 58 miles of the critical welds. All had to be ultrasonically tested to identify voids or other flaws. To make certain the ribs were airtight, the Samsung crews pumped them full of pressurized air, coated the welds with soapy water and looked for bubbles, which would indicate leaks.

Construction was further complicated by the fact that the deck tops needed to be constructed upside down, with the steel sheets on the bottom.

▲ Part of the challenge of fabricating the deck sections came from discrepancies between American and Korean welding standards. The banner in the background reminds workers that steel must be heated before and after welding—steps many Samsung welders were not used to taking.

delivered anyway. Spinning crews were able to finish the south cable with the remainder of the good wire and, as planned, they compacted that cable and hung suspenders. But spinning on the north cable didn't start up again until March. In May, when the last sections of deck were being finished on Koje, only half the suspender cables were up.

Rondón publicly downplayed the deck problems. "Large engineering and construction projects such as the new Tacoma Narrows bridge sometimes encounter challenges that require special efforts and diligence," he said. "The fabrication of the deck sections has been one of those challenges for which TNC has taken appropriate steps to keep the project on target."

But at the end of April 2006, it became clear there was no way TNC could get the bridge finished and open to traffic in the 12 months remaining before the state's deadline. Even if everything with the lifting went as well as possible, winter rains would start before the deck could be waterproofed and paved.

In early May, Rondón advised Transportation Secretary MacDonald and Linea Laird that TNC would miss its April 2 completion deadline and would aim instead for an opening day three months later, in July 2007.

MacDonald and his public relations team tried to head off criticism by putting a positive spin on the delay, referring to it in a cheerful news release not as a missed deadline but as an "updated project schedule."

"Only a little ground has been lost against our earlier hope that we would achieve an opening next April," Laird was quoted as saying in the release. "But in just a few weeks the first of the bridge deck

A WALKOUT AND A LAWSUIT

On September 5, 2005, Samsung dropped a bombshell. With costs climbing and its pleas for more money being ignored, the shipbuilder pulled its workers off the Tacoma job and curtly told the NSKB representatives at the shipyard to get out and turn in their gate passes.

Rondón and Climie, fearing the situation might hold up the bridge for months, or even years, hustled to negotiate with Samsung. NSKB declined to take part, maintaining that Samsung should be held to the agreed-on price.

In emergency sessions, TNC struck a deal with Samsung, negotiating on NSKB's behalf in its absence, for another payment beyond the agreed upon price. Samsung wanted $43 million but settled for $25.5 million.

NSKB refused to pay a share of the settlement and instead sued in Thurston County Superior Court, asking the judge to order Samsung back to work. TNC went ahead with the deal anyway, paying NSKB's share by drawing $12.9 million on a letter of credit TNC held at NSKB's bank. At the Thurston County clerk's office in Olympia, the file of documents in the lawsuit grew to thousands of pages.

It was nearly a month before workers on Koje went back on the job. By that time, it was clear that TNC's hoped-for deadlines would be impossible to meet. TNC originally had wanted the first shipment of deck sections delivered in February. In fact, the first of three shipments didn't arrive until June.

As it turned out, that didn't much matter. Back in Tacoma, the problems with the corroded cable wire had delayed cable spinning so long that TNC crews wouldn't have been ready to lift the deck sections when they were supposed to have been

◀ The delay in Korea turned out not to matter because cable work at the Narrows was running so far behind schedule. Here, steel brackets that the suspenders will hang from are hoisted into position on the main cables.

▶ The Swan arrived in Tacoma with the first 16 deck sections on June 8, 2006. Stacked four high, the sections towered 120 feet over the ship's deck.

sections will be delivered and we will be turning the corner to home."

MacDonald pointed out that in a project so complex and lengthy—55 months—a delay of three months was not unreasonable and was in fact still a remarkable achievement. "That still deserves the silver medal, or maybe the gold medal," he said.

But there was no denying the disappointment many on the job felt, from Rondón on down. Missed deadlines mean hurt pride in the construction industry.

And they also mean financial penalties. TNC's contract with the state called for penalties of $12,500 a day for the first 90 days it came in after the April deadline. If the bridge did not open by July 2, the penalty would jump to $125,000 a day, up to a maximum of $45 million.

Laird and Rondón flew to Korea in May to watch as the first 16 deck sections were loaded onto the semi-submersible Dutch transport vessel, the Swan. Stacked four high, the load topped out at 120 feet above the ship's deck.

Shortly before sunset on June 8, the Swan cruised quietly into Commencement Bay, ending an 18-day journey of 5,250 miles. The ship arrived without fanfare. A handful of people on the deck of Cliff House Restaurant watched as the ship rounded Browns Point. A Foss tug sped out in welcome, spraying a celebratory stream of water into the air and then guiding the ship to its anchorage at the head of the Hylebos Waterway.

Climie, ever the optimist, took all the delays in stride and looked forward to the lifting operation. "It will be a four-month job," he predicted.

As it turned out, that was too optimistic.

HEAVY LIFTING

After the fiasco with the corroded wire and the delays in South Korea, TNC needed things to go right with the deck-lifting operation. But the bridge builder stumbled at the gate, then hit a long patch of bumpy road.

The Swan spent two weeks in Commencement Bay, getting outfitted with lifting equipment. On June 23, when the big ship rounded Point Defiance and headed into the Tacoma Narrows with the first load of deck sections, people gathered along the shore to watch, eager to get a glimpse of history in the making.

TNC's plan was to tow the Swan into the Narrows, pass beneath old bridge and stop directly under the cables of the new bridge, about 300 feet from the Gig Harbor shore. The ship was to remain there while lifting crews plucked the 16 deck sections from it, one by one, and hung them in their proper places on the suspension cables, like clothes on a clothesline.

While the Swan was being unloaded, its nearly identical sister ship, the Teal, would make its way from Korea with a second batch of sections. When the last section had been lifted off the Swan, the Teal would take its place under the bridge and the Swan would head back to South Korea for the third and final load.

The Swan and the Teal were large ships, as long as city blocks and more than a hundred feet wide. They were able to carry such tall, heavy loads because they had enormous bulk below the water surface. That made them unusually difficult to maneuver in heavy current, especially in conditions as tight and potentially treacherous as those in the Narrows.

▲ The scale of the deck-lifting operation was difficult to comprehend without getting up close. Typical deck sections were 120 feet long and weighed 450 tons. Here, a section near the Tacoma tower is lifted into place as a barge passes below. More deck sections wait on the delivery vessel visible on the right.

Because it would be impossible to maneuver the Swan under the bridge and anchor it when the tide was running, TNC chose to take the ship to its Narrows anchorage during the hour or so between tides, when the water was relatively calm. On the chosen day—Friday, June 23—the slack tide happened to be one of the highest the year, 12.4 feet above its midpoint.

It also happened to coincide with the regular Friday afternoon commuter scramble on the old bridge.

Fearing monster traffic jams, the Department of Transportation warned drivers to keep their eyes on the road and not the ship. But the sight was hard to ignore. As the Swan entered the relatively confined space of the Narrows, the bright orange ship and green deck sections looked like impossibly large toys.

About 50 people took advantage of the warm, sunny weather and headed down to Narrows Park to watch the arrival. The park was on the Gig Harbor side, just south of the bridge, and it would have an excellent view of the ship as it passed beneath the old bridge.

Barbara and Paul Steenvoorde got to the park before anybody else. The retired Gig Harbor residents were bridge enthusiasts who had avidly followed the progress of bridge construction from the time TNC towed the cutting edges into the Narrows three years before.

Even they hadn't planned to get there as early as they did, however. Paul misread an announcement of the ship's arrival in the newspaper, and they arrived at the park with their binoculars, sun hats, and folding chairs two hours early.

The Steenvoordes weren't bothered by the wait. They chose a spot on the beach and watched as photographers set up tripods and other bridge watch-

132

ers arranged themselves on logs facing the entrance to the Narrows.

As the Swan drew closer, escorted by tugboats fore and aft, it grew from a dot to what looked like a huge floating warehouse. The deck sections towered 120 feet over the ship, making it seem as if a good shove would topple the whole thing over.

Through their binoculars the Steenvoordes could see a dozen or so people in hard hats and orange safety vests, high above the water on the old bridge.

They were engineers who had walked down from the Gig Harbor field office to watch the ship come in. Lined up along the construction walkway that ran under the roadway, they would have the best view of all. When the Swan passed under the old bridge, the tops of the decks would be directly beneath the engineers' feet.

Dave Climie and his team had calculated that when the Swan passed under the old bridge, the clearance between its load and the bottom of the old bridge would be a comfortable 36 feet. But, Climie warned, "It's going to look like a lot less."

And it did. As the ship closed in on the bridge, it seemed to fill every inch of available space. The ship's orange bow passed under the bridge, and, as the Steenvoordes watched, they saw the row of tiny figures on the walkway grow suddenly agitated. Engineers raced from side to side on the walkway, heads swiveling back and forth from the approaching ship to the bridge. Two of them turned to face the lead tugboat and waved their arms over their heads in big urgent circles.

"Something's bothering them," Paul said, the binoculars pressed to his face.

As he spoke, a deep boom echoed across the water.

"That sounded like metal on metal," Barbara said.

AN EMBARRASSED RETREAT

Somehow, despite years of planning that had involved every aspect of the Swan's arrival, the ship and its load turned out to be too tall to fit under the old bridge.

Scaffolding on the top section hit the construction walkway and toppled over onto the steel deck, which resounded like a drum.

Just in time, the tug behind the Swan reversed its engines and managed to stop the momentum of the 15 million-pound load. After what seemed like a long pause, the Swan slowly backed away from the bridge, made a wide turn and headed back to Commencement Bay.

Practically no physical damage had been done: The deck sections themselves had not hit the bridge. No one was injured, and aside from a few scratches in the paint, there was no damage to the old bridge. Up on top, the steady flow of commuters continued uninterrupted.

The psychological damage at TNC, however, was considerable.

The story about the mishap in the next morning's News Tribune was topped by a headline that screamed, "OOPS!" in letters an inch and a half tall.

A few days later the paper ran an indignant editorial that lambasted TNC, calling the collision a "numbskull moment" and likening the mistake to the missed metric conversion that crashed the $125 million Mars Climate Orbiter satellite in 1999.

People who still harbored grudges over having to pay for the bridge with tolls crowed with delight. All you'd have to do to get the right measurement, several noted, was to stand on the bridge and drop a string with a rock on the end and measure it.

One letter writer to The News Tribune said he suspected the error was "just the tip of the iceberg of additional technical problems."

▲ Framed by the old bridge tower, the Swan, a semi-submersible cargo ship loaded with the first 16 deck sections, makes its way toward its intended berth beneath the Gig Harbor side span. Despite the high tide, the load was expected to pass easily beneath the old bridge deck.

"If the contractors messed up on something this simple," he wrote, "I wonder what other errors they have made that we don't know about yet that could lead to some potentially catastrophic consequences in the future." He recommended "a full-fledged investigation of their calculations for key, critical areas of the entire project."

That would have been an enormous waste of time. The explanation for what happened was not complicated and did not indicate a pattern of carelessness.

Somehow, early in the planning process, someone had used the wrong figure for the amount of available clearance between the water and the Gig Harbor side span. That figure continued to be accepted as a given in all subsequent calculations with no one ever thinking it necessary to go out and check.

What the incident showed, more than anything else, was that for all of the infinitely precise calculations and planning the engineers had done on the project, they were, in fact, still human.

Manuel Rondón clammed up about the incident. An official TNC statement blandly stated the obvious: "During the initial deployment, technical difficulties were encountered and the decision was made to return the ship for further evaluation." Further explanation was left to Linea Laird, who was only slightly more forthcoming.

"It was an error in a calculation," Laird said through tight jaws. "An elevation was wrong. It was based on an erroneous assumption, and it just got carried on through the process."

That erroneous assumption, it later turned out, was that both of the bridge's side spans were the same distance above the water. TNC had used a measurement for the Tacoma side span, assuming the Gig Harbor side would be the same. It was not.

From an overall construction point of view, the mishap meant next to nothing. TNC cleared the equipment off the tops of the new deck sections and added ballast to the Swan, dropping it 7.5 feet lower in the water. Six days later, during a low slack tide in the middle of the night, the ship glided under the old bridge with 15 feet to spare.

The psychological damage at TNC took longer to repair. For Rondón, whose philosophy of management depended on nurturing team spirit and pride, the public humiliation was a major setback.

Rondón had worked hard to keep his teams motivated and proud. Public recognition of their accomplishments had helped buoy them. Now, for the first time in the project, they were portrayed not as heroes but buffoons.

Rondón was angry. Portraying his engineers that way was unfair, he believed. He felt hurt and betrayed.

The incident added to the pressure of being behind schedule because of the corroded wire and to the stress of the time-consuming hassles of the NSKB lawsuit, for which Rondón was having to give repeated depositions.

Unfortunately, things were going to get worse before they got better. The Swan's humiliating entrance marked the beginning of a run of bad luck and equipment failures that stalled the lifting operation for several weeks and pushed it further into the winter weather.

STALLED IN THE SIDE SPAN

To hoist the 46 deck sections into their places on the main cables, TNC had installed eight mobile lifting devices called gantry cranes. The gantries, equipped with powerful electric motors and

▲ Moments before the load of deck sections scraped the underside of the old bridge deck, forcing the Swan to make a hasty retreat back to Commencement Bay. The foul-up was an extreme embarrassment for TNC. No real damage was done, however.

▲ One of the powerful lifting devices, called gantry cranes, that lifted the deck sections from water level. The cranes operated in pairs, riding along the tops of the cables like locomotives on railroad rails.

hydraulic jacks, were designed to ride along the tops of the cables like ultra-slow-moving locomotives on railroad rails. Dangling below them were clusters of heavy woven wire rope that would be used to haul the sections upward.

The gantry cranes would operate in pairs, one set on each of the side spans and two sets in the center span. Inside control booths perched on top of the gantries, crane operators would manipulate the deck sections below them like puppet masters with marionettes, lifting them up from the water to deck level and holding them there long enough for iron-workers to attach suspender cables.

The lifting crews would not be able to assemble the deck in a linear fashion, simply starting at one end and working their way across. That would put too much stress on the cables and towers and could force the cables out of their saddles on the tower tops. Instead, they would place the deck sections in a strategically staggered sequence to balance the stress.

Even with balanced loading, Climie and his engineers calculated, the tower tops would move back and forth as much as 2.5 feet as sections were added.

Before any deck section was transferred from the water to the lifting lines, it would have to be precisely aligned with the gantries above. Otherwise, when the gantries took the weight of the section, gravity would abruptly swing it to true vertical, endangering workers and equipment below.

Calculating the right positions for the lifts would not be difficult, but keeping the deck sections in those spots long enough for workers to hook up the cables in the Narrows currents and wind presented a difficult challenge.

That had not been a problem for builders of the 1950 bridge. Its deck had been lifted into place in individual pieces, like big Tinker Toys, rather than completed sections. While the beams dangled in midair, ironworkers fastened them together by hammering bolts into predrilled holes.

Galloping Gertie's builders, though, had faced the same problem TNC now faced, although the prefabricated sections lifted to form its deck weighed only a fraction as much as the new ones. Gertie's deck was only two lanes wide and had no stiffening trusses beneath it. Each section was just 8 feet high, 50 feet long and weighed 40 tons. That was a fraction of the new deck sections, which weighed between 260 tons and 504 tons each. Even the smallest of them was three stories high and big enough for a full NBA basketball court on top, with room around the edges for fans.

In 1940, when Gertie's first deck section was lifted off the delivery barge, it swung violently. As soon as the lifting lines took the weight, the barge and the deck section lurched away from each other. Three workers slammed themselves face down on the barge to avoid being swept overboard or crushed by the swinging section. Several others saved themselves by grabbing hold of the hanging deck section and scrambling up onto it. They remained stranded there until the section stopped swinging and a tugboat operator figured it was safe to go back underneath to rescue them.

TNC had some advantages that Gertie's builders did not have. Computers, satellite positioning devices and laser surveying equipment could more accurately define the correct starting points for the lifts.

▶ **The deck lifts attracted an enthusiastic audience of spectators such as John Geubhner of Gig Harbor pictured here, who ventured out on the sidewalk of the old bridge to watch. Drivers on the bridge, who could not see what was going on, soon grew irritated with the spectators, believing they were delaying traffic.**

Holding the sections in place long enough for the cranes to take their weight still presented a problem, however. To maneuver the Swan back and forth and from side to side beneath the gantries, TNC secured the ship to anchors at its four corners, angled well in front of and behind the ship. On board, they installed powerful winches on each anchor line. By using the winches to tighten or slacken the lines, the crews could pull the ship to any position in a 500-foot circle.

Climie had estimated that, once the Swan was connected to its anchors under the cables, it would take about two weeks to make the final adjustments to the lifting apparatus and verify that all systems were operating correctly.

Instead, six weeks of fine summer weather passed, with no sign of lifting taking place and no explanation from the bridge builder. After the humiliation of the Swan's hitting the bridge, Rondón was in no mood to share his problems with the public.

But rumors began circulating. Some workers said privately that the operation had been sabotaged, that the lines attaching the Swan to its anchors could not be budged because the winch motors had been purposely ruined.

Each of the four winches was powered by four hydraulic motors. One hired technician produced photographs of several disassembled motors with loose screws and washers inside, apparently placed in hydraulic oil inlets. The screws and washers mangled the insides of the motors, stripping gears, shattering pistons, and rendering them useless.

It was true the motors were ruined. The question of whether they had been sabotaged and, if so, who might have done it, remained unresolved, or at least unpublicized. TNC made no reports to police. For TNC, the question was essentially unimportant.

Rondón had no time to get diverted by playing detective. With deadlines slipping further and further out of sight, the important thing was to keep construction moving ahead and, if necessary, sort things out later.

A NIMBLE BARGE

Climie originally had planned to lift all 46 sections directly off the delivery ships, jockeying the big vessels into position beneath the gantry cranes as needed, even in the center span.

It quickly became clear, however, that the Narrows' wind and tides made that too risky. Several months before the deck sections arrived, Climie and his engineers came up with an alternative plan, in which the delivery ships would remain under the Gig Harbor side span. The 10 sections that belonged on that side could be lifted directly from the ship's deck to the cables. The other 36 would be lifted up off the ship and then lowered onto a specially equipped barge that would take them where they needed to go and hold them there.

No ordinary barge would be nimble enough to do that. TNC needed to create a hybrid, something like a flatbed truck modified to react like a sports car.

The barge they chose as raw material was a stable, oceangoing model named the Marmac 12, originally from New Orleans. It had a flat, open deck, 250 feet long and 72 feet wide, big enough to comfortably accommodate six tennis courts.

Seattle's Foss Maritime outfitted the barge at its Ballard shipyard, equipping it with four 750-horse-power thrusters, one on each corner. The thrusters would be individually controlled by a computerized navigation system to keep the barge treading water within 3 feet of where it needed to be.

▲ The first deck lift took place on a sunny day and attracted hundreds of sightseers. The Knapp family, from Gig Harbor, made an afternoon of it, packing sandwiches, fruit and cheese.

The deck section is riding on the Marmac 12, a barge specially outfitted with powerful thrusters on each corner so it could keep the deck sections stationary in the current.

A 30-foot mast on the barge would pick up GPS signals and send a constant stream of data into a computer in an onboard control shack. The computer would take the actual location data from the GPS signals, compare it to the ideal location loaded in its memory, and issue instructions to the appropriate thrusters to make corrections, essentially functioning like a human brain issuing balance instructions to a person standing on one foot.

During lifts, the barge would have as many as a dozen people on board—the operator and oiler, a crew of technicians, and at least five ironworkers to hook up the cables sent down from the gantries.

Safety regulations prohibited workers from riding on the sections as the gantries lifted them, so the ironworkers split themselves into two teams. The team on the barge would hook up the sections to the cables lowered from the gantries, then stand back as they were cranked up to the deck level.

When a section had been raised to the proper position, another team of eight ironworkers high above on the catwalks would climb into a mesh cage called a "man basket" and drop down to the top of the suspended section.

Once there, they'd hop out, secure the deck section to its permanent suspender cables by slipping cylindrical steel pins through holes in brackets welded to the deck tops, and ride back up in the man basket.

THE MARMAC THROWS A FIT

The first lift did not take place until August 7, a full two months after the Swan had arrived from South Korea. The long delay had not cooled public interest. The weather was fine on the day of the announced lift and hundreds of spectators watched on shore, in boats, and from the sidewalk on old bridge.

The parking lot at Narrows Park was full by midafternoon and the road to it jammed with hastily parked cars. Swarms of sailboats, powerboats and kayaks darted back and forth in the Narrows. TNC's private security boats patrolled the perimeter of the work area, using flashing red lights and sirens to warn off boaters who strayed too close.

The first section was Number 24, a 116-foot-long piece that belonged right in the middle of the 2,800-foot center span. Already, crews had lifted it off the top of the stack on the Swan and loaded it onto the Marmac.

The Marmac's crew maneuvered the barge and the section under the bridge during a slack tide shortly before 5 p.m.

▲ The first deck section, aloft at last in the center of the midspan. The Washington state ferry passing by was a strange coincidence. Ferries did not normally pass through the Narrows. This one was chartered for a private event and the guests were lucky enough to witness the lift.

As the barge hovered in the temporarily calm water, workers attached the section to cables from the gantry cranes. The gantry winches took up the slack at a slow, careful pace of less than 10 inches a minute, a rate so slow spectators could see nothing happening.

At 7:45 p.m. the section at last cleared the Marmac's deck so smoothly it produced barely a shudder. Its job finished, the barge pulled away just before sunset, leaving the section suspended above the water as if by magic, its 488-ton weight distorting the graceful catenaries of the cables into a V-shape. Several hours later, ironworkers had the section in its final position and securely fastened to suspenders.

The Marmac worked well on its first outing but quickly proved too sensitive for its own good. Three days later, the barge carried the second deck section to its position in the center span, immediately east of the section already hanging from the cables. The Marmac placed the section where its computers told it to, but that position turned out not to precisely align with reality. As the weight of the deck section began to come off the barge, gravity tugged it toward plumb.

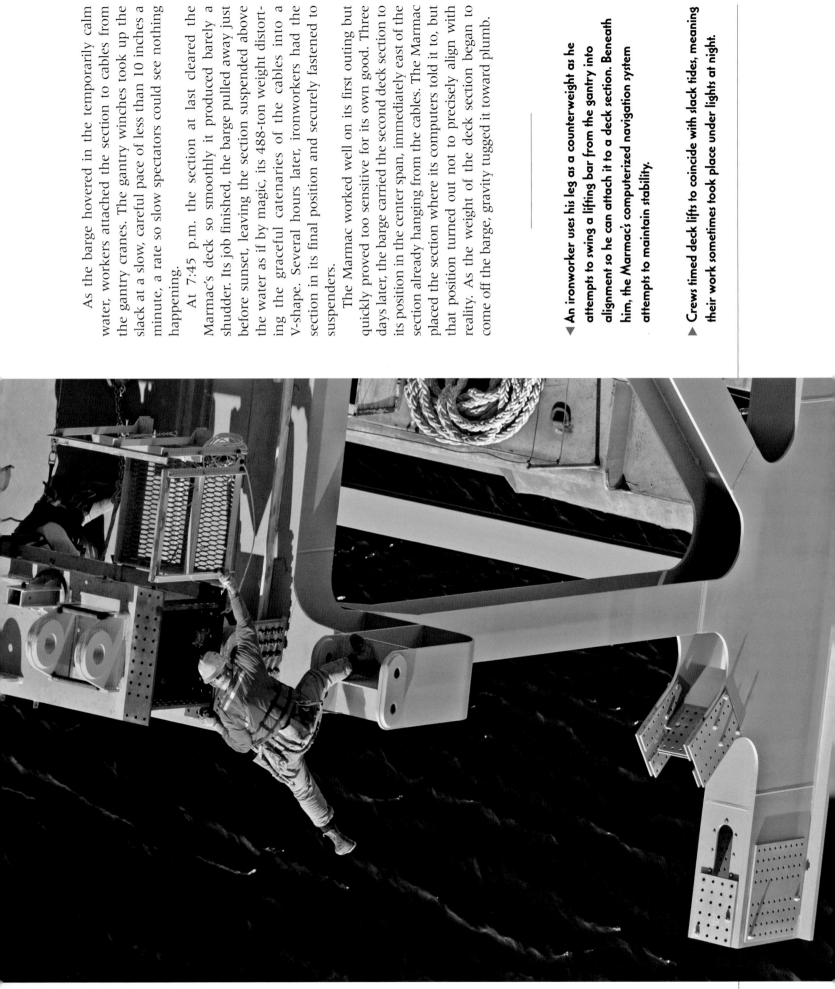

▼ An ironworker uses his leg as a counterweight as he attempts to swing a lifting bar from the gantry into alignment so he can attach it to a deck section. Beneath him, the Marmac's computerized navigation system attempts to maintain stability.

▶ Crews timed deck lifts to coincide with slack tides, meaning their work sometimes took place under lights at night.

The wind tore away wooden safety railings on the new bridge deck and broke the lines on a barge, briefly setting one of the 450-ton deck sections adrift in Commencement Bay. Trees falling on power lines knocked electricity out for more than 1 million utility customers. TNC was among those affected, losing power for most of one workday.

In January, 6 inches of snow forced workers off the bridge. Temperatures dropped so low the steel decks looked like ice cube trays. Ice coated their steel tops, making them too slippery to stand on.

Time after time, wind speeds forced TNC to shut down its construction elevators on the towers. Darkness complicated things further. Washington's gloomy latitude left only about 10 hours of light each day.

The unusually bad weather stalled the lifting operation and created problems in other aspects of the job as well. For several of the deck sections, getting them into the air was only the beginning. Sixteen of the 46 sections could not be lifted directly into place from the water because they needed to be positioned over land or between the legs of the towers. Because the gantries were unable to move along the cables when loaded, those sections, which together constituted more than a third of the 5,400-foot deck, had to be moved laterally in a precise sort of cable choreography engineers called "trapezing."

Trapezing involved two different sets of cables, one set angled in front of the other. The rear set was attached to the lifting gantries, the forward set to temporary cable bands farther down the cable.

Moving the sections sideways was a matter of sequential weight transfers, the physics of which resembled the act of walking on crutches.

It was delicate work, made more difficult by the cold and wind. On the night of January 7, four

workers who had descended 200 feet to the top of a deck section in a man basket had to spend the night there because the wind was too strong to risk bringing them back up.

Once adjacent bridge sections were hanging in the right positions on the suspension cables, their underlying trusses had to be securely bolted together, and the cold affected that, too.

Every one of the 45 joints between sections required more than 2,500 bolts, all of which needed to be inserted, capped with a washer and a nut, and tightened to precisely the right tension.

▲ Ice and snow created hazardous conditions on the deck tops, but also opportunities for fun.

▲ An unusually harsh January storm dumped 6 inches of snow on the construction site, forcing workers off the job.

Dangling 180 feet over the Narrows, workers on the MTAGs (pronounced "em-tags") essentially had no protection from the wind and cold. Bundled up in wool, polypropylene, and rain gear, they were managing to complete just one joint, on average, every five days.

"For them, the cold is the main thing that determines how effective they can be," said Tim Moore, the state's lead bridge engineer. "When they're cold, they're basically just hunched down and looking out for themselves. When the wind's blowing and you're dripping wet, you don't stuff a lot of bolts."

If possible, the welders had it worse. Their job was to join the steel deck tops together so the entire length of the bridge became a single mile-long sheet.

To connect the 46 sections, the welders needed to make 45 nearly perfect welds, each stretching across 80 feet of what eventually will be four lanes of traffic, plus pedestrian and bikeways.

Even in clear, warm weather, the job would have been technically difficult. The wind and rain complicated it enormously.

The techniques the welders used depended on keeping a layer of gas around the puddle of molten metal to protect it from atmospheric gases and to keep the electric arc stable.

If wind blew the shielding gas away, or if moisture worked its way into the mix, the weld could be compromised. When it hardened, it could be dotted with tiny craters—an unacceptable, weakened condition welders call "porosity"—or contaminated by trapped hydrogen.

As soon as they knew they'd be welding in winter weather, TNC engineers designed and commissioned four heavy-duty tents for shelter. Just designing the tents sturdy enough to hold up in the

▲ Freezing temperatures and stiff winds made conditions nearly unbearable for those in exposed positions. Ironworkers took cover when they could at this makeshift shelter on the Gig Harbor tower.

▶ Ironworkers ride a "man basket" from the catwalks down to a deck section.

Just getting to all the places that needed bolting was a major engineering feat.

TNC had four seven-person bolting crews on the bridge for two 10-hour shifts a day. Giving them places to stand while they worked took four types of specially designed moveable platforms, called MTAGs (for "moveable temporary access gantries"), that had to be swung into position and secured in and around the trusses. The bolting platforms couldn't be moved when the wind blew faster than 30 miles per hour.

Narrows took months, said Bill Madron, a TNC welding supervisor.

"There was nothing to go by," he said. "This is the first big suspension bridge built in the United States in 40 years. We've had to come from scratch on everything we're doing."

The tents, built by Yakima Tent & Awning in Yakima, stretched the entire 80-foot width of the bridge and stood about 10 feet tall, supported by frameworks of heavy steel pipes. The covering was 18-ounce black vinyl, the kind truckers use to cover loads on cross-country runs.

Once the tents were set up on the deck, access had to be tightly controlled because every time the flap opened, the wind could rush in and upset the process. Welders improved the tents' weather resistance with impromptu field modifications, including 2-by-4 reinforcements around their outside perimeters. And, to protect the welds from the drafts that found their way into the tent, even with all flaps closed, they constructed rough windshields of plywood and duct tape.

The tents were tough but not tough enough. The December 14 windstorm roared in with so much force it bent the frames, then ripped open the tents' heavy-duty vinyl sides, leaving them flapping like loose sails.

"WE HAVE A BRIDGE"

At last, in late January, the skies cleared. During the week of January 22, lifting crews put in a full six-day workweek, the first time they'd been able to do that without being forced off the bridge by one type of bad weather or another since late October.

By the middle of that week, only four gaps remained in the 5,400-foot-long deck. Those gaps

would be technically difficult to fill because they were just 8 millimeters bigger than the sections that fit into them, creating what might have been the world's biggest parallel parking problem. To make room, TNC's lifting crews had to swing several hundred feet of already joined sections out of the way, then ease the line of finished sections back to close the gaps.

The operations went forward as if charmed. By January 29, three of the four sections had been lifted into place, leaving just a single 120-foot gap about 200 feet east of the Tacoma tower. The final, missing section was already hanging from gantry lines, about 45 feet below the level of the bridge deck.

All that remained to complete the continuous span was to pull the line of eight completed sections slightly closer to the Tacoma anchorage, lift the final section into the widened gap and then ease it closed.

The Department of Transportation invited the media out on the bridge to help celebrate the final closure. Shortly after sunrise on January 30, reporters and photographers suited up in the required orange safety vests and hard hats and walked from the Tacoma anchorage onto the bridge, where they joined Secretary Doug MacDonald,

▲ Welders fusing the deck tops together had to work inside tents to keep the wind and water away from their welds.

▼ The last four deck sections presented special problems because they had to be lifted into spaces barely bigger than they were.

151

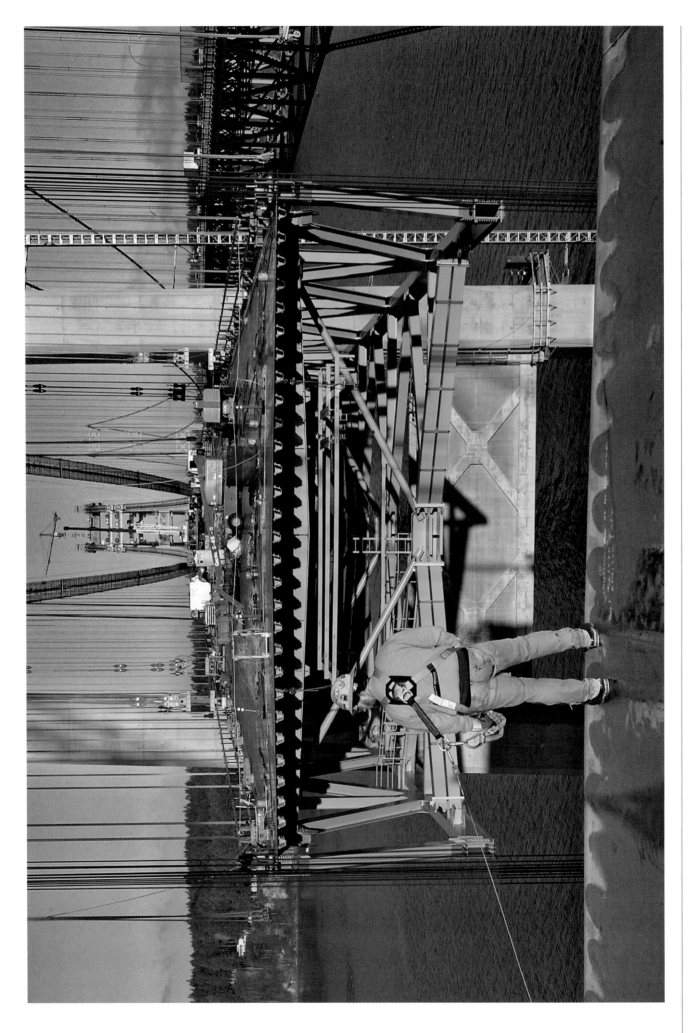

Linea Laird and Dennis Engel, the state's project engineer. Tim Moore walked out with NSKB's head of the bridge project, Kenji Sakai. A television news helicopter darted back and forth overhead.

Morning commuters on the old bridge, curious about what was going on, slowed as they passed, backing up Tacoma-bound traffic for four miles.

The weather was made to order for the occasion. The entire Olympic Range showed up for the first time in months, its heavy topping of white snow etched into the blue sky. The wind cooperated, too, fading to the slightest of breezes, a dramatic change from conditions during much of the lift operation.

When all was ready, the winch operators switched the power on and, in just 10 minutes, the last 504-ton steel structure levitated into place.

The piece fit as neatly as the final piece in a jigsaw puzzle, completing a smooth, continuous flow of steel from one side of the Narrows to the other.

A few minutes after nine o'clock, ironworkers hopped over to connect the pins to the suspender cables. With a minimum of fuss, they slipped the pins into their sockets, bringing the bridge project to a symbolic end.

Photographers in helicopters overhead shot the first pictures of, not one, but two Tacoma Narrows bridges.

An ironworker named Steve Seidel, who had worked on the bridge project all winter, strode off the section, a big grin on his face, and grabbed MacDonald's outstretched hand.

"We have a bridge," Seidel said.

MacDonald grinned back, even bigger. He thumped Seidel on the back hard enough to raise a small cloud of dust.

"Yes, we do," he said.

At that point, after more than four years of construction and 2.8 million man-hours of labor, it was possible for the first time to walk on the new span from one side of the Narrows to the other.

MacDonald pulled out his camera, herded the ironworkers into a tight group and snapped pictures. Then he handed off his camera and posed with them.

▲ Ironworker foreman Bryan Mosby keeps watch as the last deck section rises into position near the Tacoma tower.

▼ "We have a bridge." Ironworker Steve Seidel, left, celebrates the final lift with Transportation Secretary Doug MacDonald.

PRIVATE ASSURANCE

Rondón skipped the hoopla, saying he preferred to let the workers who had suffered through the winter get the recognition they deserved.

Climie didn't show for the same reason, plus, he didn't regard lifting the last deck section as the real ending.

Technically, Climie noted, the deck was still not completely joined. The eight sections closest to the Tacoma anchorage still needed to be released and eased back into their proper positions, closing a 25-inch gap bridged by an aluminum gangplank. The gap would stay there until late in the day or perhaps the next.

Climie was prepared to make the necessary corrections if the ends didn't match up when the gap closed. "This can be the fiddly part," he warned. If the sections were slightly different heights, he could adjust them with giant steel braces and wedges. If they were twisted, he was ready with steel cables and winches.

Late that Tuesday afternoon, Climie walked out on the bridge. He talked casually to workers and foremen, checking things out but, as always, keeping a low profile.

"I don't want them to think I'm looking over their shoulders," he said, "because I'm not."

When he left, he made sure the foremen had his cell phone number and told them not to hesitate to use it, just in case any unforeseen complications should come up.

Climie walked off the bridge but didn't go home. He stayed in the construction parking lot by the Tacoma anchorage and waited there until the gap was closed. His phone didn't ring.

At sunup the next morning, Climie again parked at the Tacoma anchorage and walked out onto the deck. Now that all 46 deck sections were connected, the bridge would act like a single structure for the first time.

Climie knew that for every 10 degrees of temperature change, the mile of steel would shrink or expand slightly more than 4 inches. If all had gone as projected, the bridge would be exactly 5,400 feet long at 64 degrees Fahrenheit, and, at that temperature, the center of a Teflon pad on the side of the bridge deck should be exactly in the center of a stainless steel rubbing plate on the inside of the tower leg.

The temperature that morning was about 45 degrees, enough to shrink the bridge about 8 inches. That meant the center of the pad at the Tacoma tower should be slightly to the right of dead center.

Climie walked straight to the tower and leaned over the wooden safety railing to check out the alignment.

He didn't have a ruler, but he didn't need one to see that the deck was exactly where it should be.

He looked up smiling. It was a fine day.

▲ The finished span as seen from a helicopter on the day after the last gap was filled. After months of hardship and setbacks, a fine fit.

The slight distortion in the deck's arch would smooth out when the sections were permanently connected and the heavy lifting equipment removed from the main cables.

FINISHING UP

TNC puts the finishing touches on the bridge, getting set for opening day. The dream team breaks up—and some take a step back to consider what they have created.

Two hours before sunrise on February 5, 2007, pickup trucks from all over the South Sound converged on TNC's Gig Harbor field office. By 5:30 a.m. the parking lot was full, and drivers were jockeying for parking spaces on the shoulder of surrounding streets.

It was the first Monday of the month and time for TNC's regular all-staff meeting.

A hundred ironworkers, laborers and equipment operators in boots and hard hats made their way through the dark to the front door of the office, where a raised wooden deck served as an outdoor speakers' platform. They stood waiting, hands in their pockets for warmth or drinking coffee from cardboard cups, exhaling steam into the cold air.

The early morning meetings, standard practice since the bridge project began, gave Manuel Rondón and his top managers a chance to reiterate the need for safety, quash rumors, and, when necessary, renew enthusiasm.

This Monday meeting, however, was not going to be a typical one. Off to one side of the building, fluorescent office lights reflected off the hood of a dark blue limousine, parking lights on and motor idling.

At 6 o'clock sharp, Pat Soderberg, TNC's big, laconic construction manager, stepped out of the office with a wireless microphone in his hand. He started out by congratulating everybody for safely finishing the bridge deck.

▲ **And then there were two. A foggy morning sunrise behind Mount Rainier illuminates the new profile of the Narrows bridges. The new span was complete, but months of finishing up remained before opening day. Among the first tasks: dismantling the gantry cranes.**

When the deck was finished, Governor Chris Gregoire paid a surprise visit to the TNC field office to congratulate workers on a job well done. Here, she hugs former State Senator Bob Oke, who pushed hard for a second bridge when he was in the Legislature. He died weeks before the bridge opened. In the background, both in safety vests, are Tim Moore, the Transportation Department's top bridge expert on the left and Transportation Secretary Doug MacDonald on the right.

"Give yourselves a round of applause," he told them.

A few workers clapped self-consciously. Most just looked around and grinned.

Then, a surprise visitor: Governor Chris Gregoire stepped out of the shadows and onto the lighted platform, vivid in a yellow rain jacket. The microphone had mysteriously gone dead, so she had to shout to be heard over the noise of Highway 16, roaring to life with morning commuters.

"It's a beautiful bridge," Gregoire told the gathered workers. "You deserve to be proud of yourselves.

"I travel all over the world on trade missions, and I can tell you that I put American workers right up there with the best in the world. You will be part of history."

The governor's congratulations were appreciated, but slightly premature. The shore-to-shore connection had been established across the Narrows, but a fair amount of work remained before the bridge could technically be called finished.

TNC crews still had to wrap and paint the main suspension cables, and they had to pave the new deck—both time-consuming tasks that would be complicated by heavy spring rain. They had to install 100-ton roadway expansion joints at both ends of the bridge, the first of which would end up stalled at the Washington–Idaho border for three weeks because the trucking company ran afoul of the state's weight requirements. They had to build traffic barriers, set up pedestrian railings, paint stripes on the lanes, and see to a thousand other details.

But at that point, the spectacular engineering feats already had been accomplished and, in the greater context of the five-year project, the work that remained amounted to little more than finishing up.

The governor was not the only one whose thoughts had moved forward to opening day and beyond. At the state Department of Transportation, Doug MacDonald and Linea Laird were well into plans for a grand-opening ceremony. After reviewing proposals from a half-dozen event-planning firms, they had hired a Seattle company called The Workshop, which had handled the grand openings of the Seattle Mariners' baseball stadium and the Seattle Symphony's concert hall, and organized annual New Year's Eve fireworks shows at the Space Needle.

MacDonald was adamant about wanting a "people's ceremony," one that would honor the workers and give ordinary citizens a chance to walk across the new bridge before it opened to traffic. But at the same time, neither he nor anybody else wanted to repeat the mistakes San Francisco had made in 1987 at the 50th anniversary of the Golden Gate Bridge.

There, an estimated 800,000 people—16 times as many as expected—turned out for "Bridgewalk '87." People poured onto the bridge from both ends, walking, on roller skates, pushing baby strollers and even on stilts. Together, they added so much weight to the bridge that they temporarily lowered its arch by 10 vertical feet.

For more than an hour, people were packed so tightly in the middle of the bridge that no one could move. Then-mayor of San Francisco, Dianne Feinstein, gave up on delivering the speech she had planned and managed to escape only after her staff linked arms in a flying wedge formation and propelled her through the crush of bodies.

To avoid such a scene, MacDonald and Laird decided instead on a "progressive opening," with not just one ribbon-cutting ceremony but several spread

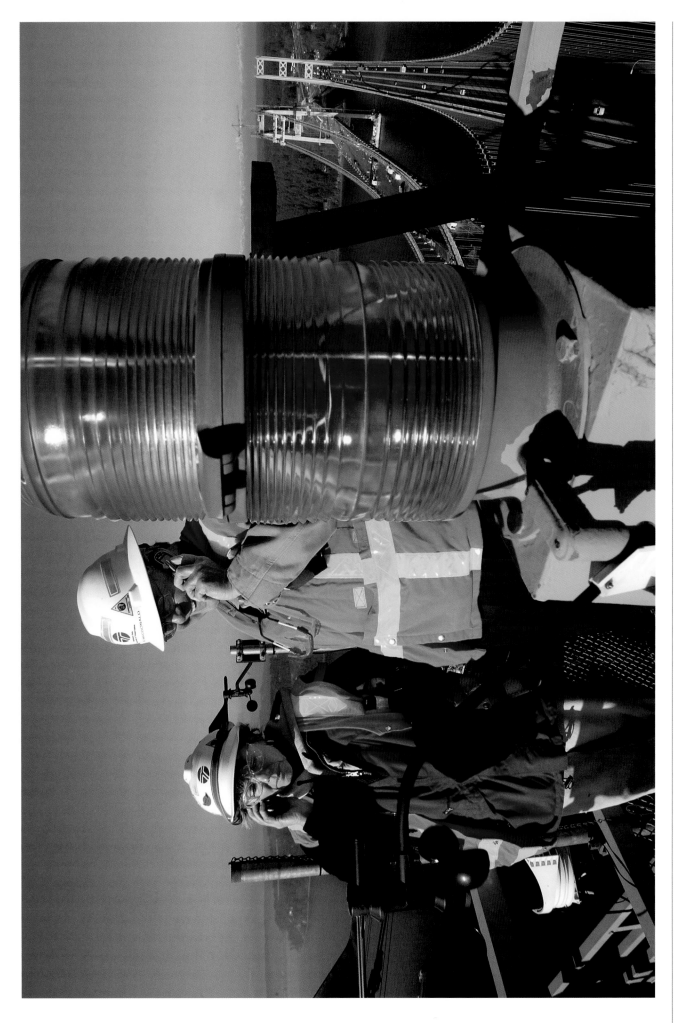

over 12 hours. They wanted to go easy on the VIPs and self-congratulatory speeches, offering instead an opportunity to stroll leisurely across the bridge, admiring the structure and the views of Mount Rainier and the Olympics. A lottery among bridge workers would determine who would cut the ribbons. Shuttle buses would pick people up at designated gathering points and haul them to the bridge and back throughout the day.

Their plans for the ceremony grew more modest after word got out that whatever it cost would have to be repaid with bridge tolls. "It's money that should be used for potholes, not parties," a Republican senator from the eastern side of the state grumped.

Another celebration began months before the bridge opened. In April, the Washington State History Museum devoted most of the exhibit space in its downtown Tacoma museum to a commemorative "Bridging the Narrows" exhibition. The eight-month show celebrated not just the new bridge but its two predecessors as well. Visitors walked across a re-created catwalk to enter the exhibit and were able to touch a piece of roadway and curb from the wreckage of Galloping Gertie.

And, as TNC crews put the finishing touches on the bridge, a group of residents turned up the juice on a grass-roots campaign to put permanent arrays of decorative lighting on both bridges.

The lighting advocates, inspired by the work lights that lit up the catwalks during construction, lobbied members of the state Legislature, local city councils and community service groups, looking for the $4 million they estimated it would cost to illuminate the bridges with energy efficient LED fixtures powered by solar panels.

In addition to being beautiful, they said, the lights would produce intangible marketing benefits.

The parallel bridges would become iconic branding symbols for the area, like Seattle's Space Needle, St. Louis' Arch and the Eiffel Tower in Paris.

THE TEAMS BREAK UP

As the technical work on the bridge wound down, the carefully assembled teams of engineers, ironworkers and managers began breaking up, their members moving on to other jobs.

It was a pattern most in the construction industry were used to. The first thing ironworker foreman Bryan Mosby said after attaching the last deck section and shaking Secretary MacDonald's hand was, "OK, we're done. Time to start looking for work elsewhere."

As it turned out, MacDonald would be moving on as well. In April 2007, he unexpectedly announced that he intended to retire as secretary of transportation almost immediately after the bridge opened.

For most of the tradesmen, heading for jobs on office buildings, shopping centers and warehouses, the bridge job would be the high point of their careers, both in terms of technical difficulty and bragging rights.

Except for top specialists like Dave Climie and a handful of others, the chances of ever working on another suspension bridge were remote. No other suspension bridges were under construction anywhere else in the world in the spring of 2007. In the United States, only four had been built in the past half-century, and no others were in serious planning stages.

Adrianne Moore, the Tacoma woman who won her job wrestling concrete for the bridge's caissons and towers by showing up early at the job site and handing out doughnuts, took six months off to recover physically after her bridge job ended.

▲ **From the old bridge's Tacoma tower, the state's project manager, Linea Laird and Secretary Doug MacDonald whip out cell phones to let friends and colleagues know the bridge is finished.**

Transportation Department officials had to keep plans for the opening-day celebration modest for political reasons: The money for the celebration would need to be paid back with bridge tolls.

there. I was awestruck every single day. I will be proud of this for the rest of my life."

Top engineers and managers at TNC and in the state's project office cleaned out their desks and moved on, their specialized skills too valuable to waste on the relatively straightforward work that remained on the bridge.

Tom Sherman, who supervised the construction of the caissons supporting the towers, was already retired, having left the massive bridge foundations as his final project. He did some consulting work on marine projects and once got dressed up in what he called his "penguin suit" to accept the Heavy Engineering Construction Association's Golden Beaver award in Los Angeles for his work on the bridge. Most of his time, though, he spent working on his waterfront property on Vashon Island and puttering with his collection of antique cars.

The contractual obligations of TNC's Japanese partners, Nippon Steel and Kawada Bridge, ended with the completion of the superstructure, and Kenji Sakai and his team of suspension bridge experts gradually rotated back to Japan and assignments on other projects.

Tim Moore, the state's bridge specialist who had been on the Narrows project longer than any other state employee, took a position as a member of the Department of Transportation's "mega-bridge" team in Olympia.

Climie's work ended with the completion of the superstructure, too. He sold his house in University Place, south of Tacoma, and got ready to move on. He wasn't sure where his next project would take him, but he had his eye on a half-dozen new suspension bridges being planned in other countries. China had two; Chile was in the midst of the final cost review

"I spent two years out there beating my body pretty gosh darn good," she said, recalling in particular one 21-hour day of pouring concrete. After her break she went to work as a union laborer on a new science building being built at Tacoma Community College.

Moore looked back with gratitude for the opportunity she had to work on the bridge. "The pride was monumental that I had for the job," she said. "I learned so much. The very best of the best were

▼ As work on the bridge wound down, the elite teams of builders began to break up. Concrete worker Adrianne Moore took six months off to recover physically.

on a two-span, 1.5-mile bridge across the Chacao Channel, and Norway's Hardangerfjord bridge, which would have the world's seventh largest main span at 4,298 feet, was about to go to bid.

Laird won a prestigious promotion and moved out of her Gig Harbor office to the state construction engineer's position in Olympia. Her successor, Jeff Carpenter, was to stay on as project director past the grand opening of the new Narrows bridge and a year-long retrofit of the old one. But as opening day neared, his time was increasingly spent trying to head off the traffic tie-ups feared when the new bridge opened. Immediately after opening day, the old bridge was to be closed for a year for repairs.

Carpenter and what was left of the project team were heavily pushing an automatic toll-collection system called "Good to Go!" in which people bought transponders that would charge tolls automatically and ease congestion. In the weeks leading up to opening day, a giant banner strung across the toll plaza advised passing drivers: "Skip the toll booths. Get Good to Go!"

THE PENINSULA FEELS THE PAIN

In Gig Harbor, anticipation of the bridge's opening was cooled by the realization that, immediately after the celebration, the tollbooths would open for business, extracting hard cash for every crossing.

People in Gig Harbor still complained about having to pay for the bridge with tolls, but the steam had gone out of their protests, in part because they no longer seemed to be so unfairly singled out.

Rather than raising taxes, the state Legislature seemed headed toward using tolls to finance other multimillion-dollar transportation projects. A new floating bridge across Seattle's Lake Washington and

a project to widen Interstate 90 over Snoqualmie Pass appeared to be likely candidates for pay-as-you-go plans.

After putting up with construction for five years, commuters were eager to see the new bridge open, but many were skeptical about how much it would ease congestion, particularly since the 1950 bridge would be tied up for retrofitting and repair.

The state's traffic analysts, meanwhile, had changed their predictions about how many people were likely to use the bridges. In 2000, when the decision was made to build the new bridge, an average of 88,000 vehicles were crowding across the Narrows each day. At that point, planners were predicting 120,000 daily users by 2020. As the bridge

▲ Tom Sherman, who directed work on the caissons, retired to his home on Vashon Island. He spent his spare time working on his collection of classic cars.

▲ From the top of the old Tacoma bridge tower, a 360-degree view illustrates how the Narrows controlled urban growth. Tacoma appears on the left in the photograph; the lightly developed west side is on the right. While many peninsula residents looked forward to the easier access the new bridge would provide, some dreaded the changes it would bring.

The photo above is a digital compilation of two-dozen images.

neared completion, however, the planners lowered their forecasts, citing slower-than-expected employment and housing growth.

The toll would discourage use of the bridge, too, they said, reducing the number of cars and trucks by about 15 percent. All things considered, traffic was now expected to increase only to about 95,000 cars by 2020.

If there had been a slowdown in growth, it was not apparent in Gig Harbor. The economy on the west side had boomed throughout the years of bridge construction, with new subdivisions, shopping centers, medical centers, and chain stores popping up throughout the traffic corridor. One single development called Harbor Hill, planned for 320 acres just north of Gig Harbor, envisioned a thousand new homes, crowned by a 150,000-square-foot Costco store.

In 1940 and 1950, when the other two Narrows bridges had opened, west side residents had nearly unanimously applauded them for the access they would provide. In 2007, a considerable number of people looked at the new bridge with dread, equating easier access with urban sprawl.

Those who cherished the peninsula's isolation feared it would be impossible to keep urban crime and ugliness from spilling across the bridge and destroying what many of them had moved to the west side to escape.

Pat Lantz, the Gig Harbor area's representative in the state House of Representatives, took a field trip to the top of the Gig Harbor tower shortly before the bridge opened and was stunned by the view back toward her home.

The new bridge's toll plaza and western approaches had transformed the landscape on the Gig Harbor side. More than $1 million worth of landscaping, which Lantz had successfully lobbied for, could not disguise the fact that what just five years before had been a four-lane road tunneling through towering fir trees, was now a hard and inhumane expanse of concrete.

"To look west is to be speechless," Lantz said. "We have a community that has been totally transformed."

From the bridge tower, Lantz had a clear and disturbing vision of the future: freeways, subdivisions, and shopping malls stretching from Gig Harbor all the way to Bremerton.

Because of the invitation presented by the new bridge, she said, "The scale of everything has been ratcheted up, up, and up. Before long, there's not going to be any memory of what we once had."

But where some saw encroaching ugliness, others saw jobs and prosperity. Retailers and developers in peninsula towns all the way north to Port Townsend, Sequim, and Port Angeles looked forward to a rush of investment when the new bridge widened the bottleneck at the Narrows.

Bob Oke, the Port Orchard Republican who had been the new bridge's most vigorous champion in the state Senate through most of the 1990s, was diagnosed with a rare form of blood cancer while the bridge was being built and retired from public office in 2006.

As opening day approached, he was delighted with the new bridge and the economic development it was encouraging. Despite an exhausting regimen of medical treatments, he was busily trying to convince his former colleagues in government to subsidize an 80,000-seat NASCAR racetrack in rural Kitsap County that would draw thousands of spectators across the bridge. The project ultimately failed for lack of support.

"This area has been neglected for years and years by big business," Oke said. "We have a huge potential for industrial development and good paying jobs. Businesses love the area but they've always said, 'How do I get my goods and services in and out?' Now we've solved that problem."

Oke never got to see the bridge open. He died on May 14, 2007.

RONDÓN REFLECTS

Manuel Rondón would stay on the job as project manager, at least through opening day. After that, he didn't know, or at least wasn't saying.

A few weeks before the bridge opened, Rondón took a break from the TNC office and drove down to the beach at Narrows Park, his favorite place for contemplation and decompression.

During particularly stressful days in the office Rondón sometimes sent frazzled engineers to the little park to gaze out at the bridges from a distance and regain a sense of calm and perspective.

The park had changed dramatically in the five years he had been going there. Before construction

▲ West of the bridges, the wide swath of concrete that replaced acres of forest. "To look west is to be speechless," state Representative Pat Lantz said sadly. "We have a community that has been totally transformed."

began it had been a local secret. The way to the beach, a rutted dirt trail, was used mostly by fishermen.

Now, the blackberry thickets were gone, replaced by a 25-car parking lot. The trail to the beach had been transformed into a concrete sidewalk with steel handrails. Bridge gazing had been helped along with two telescopic viewers, courtesy of TNC.

The real change, though, was the view across the water. The old bridge profile, a single fragile ribbon across the water, was now partially obscured and complicated by the square, solid towers of the new bridge.

At that point, Rondón was into his seventh year on the Narrows job. He had celebrated his 50th birth-

day on the project, and his children had grown up during it. He was heavier, his beard grayer. He sat on a log on the beach and looked at the bridges. Barely visible, near the top of the new bridge, workers were painting the cables green to match the old ones.

Five years earlier, on this same beach, Rondón had laid out the criteria he would use to judge whether the bridge project had been a success. Had he maintained a safe working environment? Did it come in on time and within budget? Did the quality meet the client's expectations?

With regard to safety, the project had been an unquestioned success. Rondón's biggest dread, that a worker would be killed during his watch, had been avoided. After more than 3 million hours worked, there had been only a few sprains, broken bones, pulled muscles and smashed thumbs.

The quality of workmanship and materials was also unquestioned. State Transportation Department inspectors were delighted with the results.

As for coming in within budget, the state had a guaranteed price from the beginning, so that was never in question. Rondón would not discuss how much Bechtel and Kiewit, TNC's parent companies, were likely to make on the project, but others in his organization said the construction giants would make plenty—though not as much as they had hoped.

The corporate profits depended in part on the resolution of fault regarding the corroded wire and the outcome of the lawsuit over the bridge decks with Nippon Steel and Kawada Bridge, set for trial in March 2008.

The project's most obvious shortcoming was the matter of time. TNC missed its April 2007 completion deadline. Delays in South Korea, the corroded wire and the problems with the winch motors on the deck-delivery ships set the project back three months.

But, on a job scheduled to last almost five years, a delay so short was still a respectable finish, Rondón said.

"I believe that we did a good job in managing incredible adversity in this job that we never anticipated in the beginning," he said. "To manage a situation like that and to have kept the delay as short as we did, I believe that is a success story."

For Rondón, in a reflective mood, those easily quantifiable measures did not begin to tell the whole story. Success is relative, he said. "If you're too hard on yourself, you never succeed. If you're too easy on yourself, you never fail."

The important thing was the fact of the bridge itself, out across the water. He pointed out the precisely parallel arches of the roadways, the way the towers from different eras complemented one another.

"It looks good," he said. "It looks fantastic."

As important as the structure, Rondón said, were the experiences and the memories of the experiences, secure in the minds of hundreds of workers who had pushed themselves to their limits on the job, and now could look back with pride.

Rondón counted himself among them. "It is a project in which I have invested probably the best years of my life," he said. "I was able to take it in its true infancy—from preconception really—and I was immersed in every step along the way.

"I've had seven years of a great job, working with great people in a great community. I look back and see now what wonderful experiences they were. We have been in paradise here."

▲ Manuel Rondón at Narrows Park, seven years after taking on the job as TNC's project manager. "The real fun is in doing," he said. "It's not in finishing."

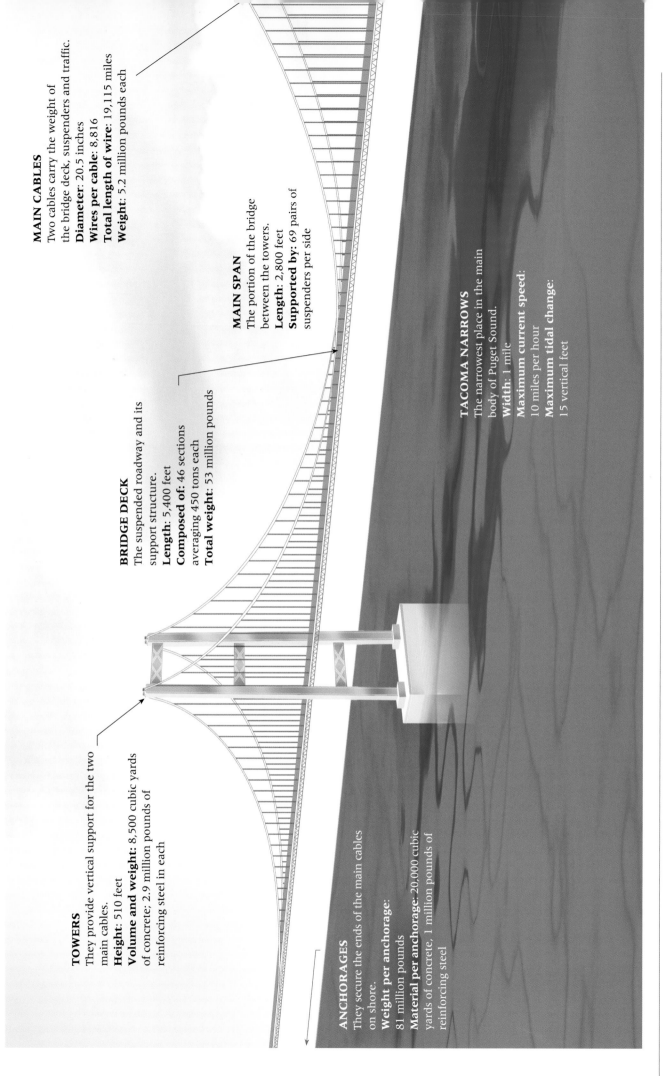

MAIN CABLES

Two cables carry the weight of the bridge deck, suspenders and traffic.
Diameter: 20.5 inches
Wires per cable: 8,816
Total length of wire: 19,115 miles
Weight: 5.2 million pounds each

MAIN SPAN

The portion of the bridge between the towers.
Length: 2,800 feet
Supported by: 69 pairs of suspenders per side

BRIDGE DECK

The suspended roadway and its support structure.
Length: 5,400 feet
Composed of: 46 sections averaging 450 tons each
Total weight: 53 million pounds

TACOMA NARROWS

The narrowest place in the main body of Puget Sound.
Width: 1 mile
Maximum current speed: 10 miles per hour
Maximum tidal change: 15 vertical feet

TOWERS

They provide vertical support for the two main cables.
Height: 510 feet
Volume and weight: 8,500 cubic yards of concrete, 2.9 million pounds of reinforcing steel in each

ANCHORAGES

They secure the ends of the main cables on shore.
Weight per anchorage: 81 million pounds
Material per anchorage: 20,000 cubic yards of concrete, 1 million pounds of reinforcing steel

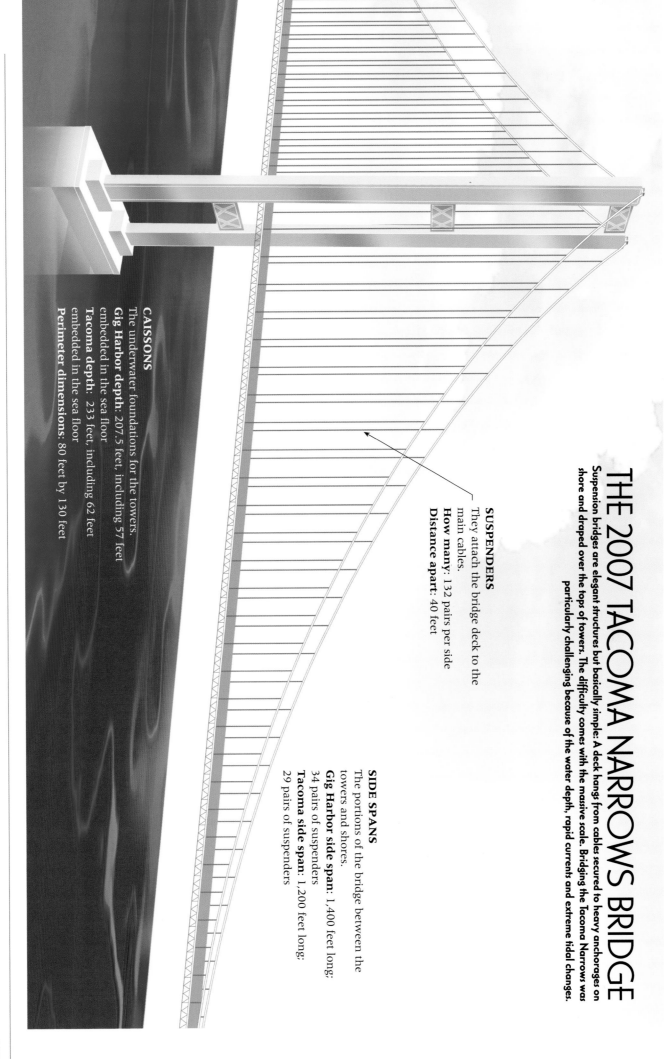

THE 2007 TACOMA NARROWS BRIDGE

Suspension bridges are elegant structures but basically simple: A deck hangs from cables secured to heavy anchorages on shore and draped over the tops of towers. The difficulty comes with the massive scale. Bridging the Tacoma Narrows was particularly challenging because of the water depth, rapid currents and extreme tidal changes.

SUSPENDERS

They attach the bridge deck to the main cables.

How many: 132 pairs per side

Distance apart: 40 feet

SIDE SPANS

The portions of the bridge between the towers and shores.

Gig Harbor side span: 1,400 feet long; 34 pairs of suspenders

Tacoma side span: 1,200 feet long; 29 pairs of suspenders

CAISSONS

The underwater foundations for the towers.

Gig Harbor depth: 207.5 feet, including 57 feet embedded in the sea floor

Tacoma depth: 233 feet, including 62 feet embedded in the sea floor

Perimeter dimensions: 80 feet by 130 feet

ABOUT THE AUTHOR

Rob Carson is a special projects reporter at *The News Tribune* in Tacoma. He was a finalist for the Pulitzer Prize for explanatory writing in 1992 and spent nearly five years documenting the construction of the new Tacoma Narrows bridge. Carson is also the author of *Mount St. Helens: The Eruption and Recovery of a Volcano*, and *The Living Mountain*.

ABOUT THE PHOTOGRAPHER

Dean J. Koepfler has been a staff photographer at *The News Tribune* since 1989. He was the National Press Photographers Association's regional Photographer of the Year in 1996. His work in China in 2005 won a first place award from the Society of Professional Journalists and an International Perspective Award from the Associated Press Managing Editors association.

SPECIAL THANKS TO:

Earl White

Tim Moore

Dave Climie

Dennis Engel

Claudia Cornish

Erin Babbo-Hunter

Dave Davis

Larry and Marilyn Bielstien

Gary and Sue Petersen

Robert Mester

Randall McCarthy

Jeremy Harrison

Bettie Lee Koepfler

Liz, Jordan and Evan Koepfler

Lyn Smallwood Carson

Theodora Carson

Dedicated to the memory of

Roy Anderson Gallop.

Graphic artist, bridge enthusiast, friend.

Roy fell in love with suspension bridges while working on this project.

He crossed a bigger bridge on May 25, 2005.

PHOTO CREDITS

All photos by Dean J. Koepfler except the following:

Page 7, Gig Harbor Peninsula Historical Society, Photo by James Bashford

Page 9, Gig Harbor Peninsula Historical Society, Photo by Joe Gotchy

Page 10, Tacoma Public Library, Richards Studio Collection D32519-1

Page 12, Tacoma Public Library, Lee Merrill

Page 13, Gig Harbor Peninsula Historical Society, Photo by James Bashford

Page 14, Tacoma Public Library, Richards Studio Collection D9944-15

Page 16, University of Washington Libraries, Special Collections, PH Coll. 290.33b, 290.33f

Page 17, The News Tribune, James Bashford

Page 18, University of Washington Libraries, Special Collections, UW21414

Page 19, The News Tribune

Page 20, (top) Gig Harbor Peninsula Historical Society, Photo by Joe Gotchy

Page 20, (bottom) Courtesy of Earl White

Page 23, The News Tribune, Geff Hinds

Page 36, Courtesy of Earl White

Page 46, Underwater Admiralty Services, Inc., John Schliemann

Page 47, Underwater Admiralty Services, Inc., John Schliemann

Pages 48, 49, The News Tribune, Lui Kit Wong

Page 63, The News Tribune, Drew Perine

Pages 108, 109, The News Tribune, Bruce Kellman

Page 171, The News Tribune, Janet Jensen

THE NEWS TRIBUNE

www.thenewstribune.com

© 2007 by Tacoma News, Inc. All rights reserved.
No part of this work may be reproduced, or transmitted in any form by any means, electronic or mechanical, including photocopying and recording, or by any information storage or retrieval system, without the permission in writing from the publisher.

Copy editing: Cole Cosgrove
Book design: Constance Bollen, cbgraphics
Cover design: Fred Matamoros
Production management: Nancy Duncan

Library of Congress Control Number: 2007927890

ISNB: 978-0-9633035-1-6
0-9633035-1-1

Printed in China